100
Structure and Bonding

Editorial Board:
A.J. Bard · I.G. Dance · P. Day · J.A. Ibers · T. Kunitake
T.J. Meyer · D.M.P. Mingos · H.W. Roesky
J.-P. Sauvage · A. Simon · F. Wudl

Springer
Berlin
Heidelberg
New York
Barcelona
Hong Kong
London
Milan
Paris
Singapore
Tokyo

π-Electron Magnetism
From Molecules to Magnetic Materials

Volume Editor: J. Veciana

With contributions by
D. Arčon, M. Deumal, K. Inoue, M. Kinoshita,
J.J. Novoa, F. Palacio, K. Prassides,
J.M. Rawson, C. Rovira

 Springer

The series *Structure and Bonding* publishes critical reviews on topics of research concerned with chemical structure and bonding. The scope of the series spans the entire Periodic Table. It focuses attention on new and developing areas of modern structural and theoretical chemistry such as nanostructures, molecular electronics, designed molecular solids, surfaces, metal clusters and supramolecular structures. Physical and spectroscopic techniques used to determine, examine and model structures fall within the purview of Structure and Bonding to the extent that the focus is on the scientific results obtained and not on specialist information concerning the techniques themselves. Issues associated with the development of bonding models and generalizations that illuminate the reactivity pathways and rates of chemical processes are also relevant.

As a rule, contributions are specially commissioned. The editors and publishers will, however, always be pleased to receive suggestions and supplementary information. Papers are accepted for *Structure and Bonding* in English.

In references *Structure and Bonding* is abbreviated *Struct Bond* and is cited as a journal.

Springer WWW home page: http://www.springer.de
Visit the SB home page at http://link.springer.de/series/sb/ or
http://link.springer-ny.com/series/sb/

ISSN 0081-5993
ISBN 3-540-41680-3
Springer-Verlag Berlin Heidelberg New York

CIP Data applied for

This work is subject to copyright. All rights are reserved, whether the whole or part of the material is concerned, specifically the rights of translation, reprinting, reuse of illustrations, recitation, broadcasting, reproduction on microfilm or in any other way, and storage in data banks. Duplication of this publication or parts thereof is permitted only under the provisions of the German Copyright Law of September 9, 1965, in its current version, and permission for use must always be obtained from Springer-Verlag. Violations are liable for prosecution under the German Copyright Law.

Springer-Verlag Berlin Heidelberg New York a member of BertelsmannSpringer
Science + Business Media GmbH
http://www.springer.de
© Springer-Verlag Berlin Heidelberg 2001
Printed in Germany

The use of registered names, trademarks, etc. in this publication does not imply, even in the absence of a specific statement, that such names are exempt from the relevant protective laws and regulations and therefore free for general use.

Typesetting: Scientific Publishing Services (P) Ltd, Madras
Production editor: Christiane Messerschmidt, Rheinau
Cover: Medio V. Leins, Berlin
Printed on acid-free paper   SPIN: 107 023 03   02/3020 – 5 4 3 2 1 0

Volume Editor

Professor Jaume Veciana
Campus Universitari de Bellaterra
Institut de Ciència de Materials de Barcelona
01893 Cerdanyola
Spain
E-mail: vecianaj@icmab.es

Editorial Board

Prof. Allen J. Bard
Department of Chemistry and Biochemistry
University of Texas
24th Street and Speedway
Austin, Texas 78712, USA
E-mail: ajbard@mail.utexas.edu

Prof. Peter Day, FRS
Director and Fullerian Professor of Chemistry
The Royal Institution of Great Britain
21 Albemarle Street
London WIX 4BS, UK
E-mail: pday@ri.ac.uk

Prof. Toyohi Kunitake
Faculty of Engineering:
Department of Organic Synthesis
Kyushu University
Hakozaki 6-10-1, Higashi-ku
Fukuoka 812, Japan
E-mail: kunitcm@box.nc.kyushu-u.ac.jp

Prof. D. Michael P. Mingos
Principal
St. Edmund Hall
Oxford OX1 4AR, UK
E-mail: michael.mingos@seh.ox.ac.uk

Prof. Jean-Pierre Sauvage
Faculté de Chimie Laboratoires de Chimie
Organo-Minérale
Université Louis Pasteur
4, rue Blaise Pascal
67070 Strasbourg Cedex, France
E-mail: sauvage@chimie.u-strasbg.fr

Prof. Fred Wudl
Department of Chemistry
University of California
LosAngeles, CA 90024-1569, USA
E-mail: wudl@chem.ucla.edu

Prof. Ian G. Dance
Department of Inorganic and Nuclear Chemistry
School of Chemistry
University of New South Wales
Sydney, NSW 2052, Australia
E-mail: i.dance@unsw.edu.au

Prof. James A. Ibers
Department of Chemistry
North Western University
2145 Sheridan Road
Evanston, Illinois 60208-3113, USA
E-mail: ibers@chem.nwu.edu

Prof. Thomas J. Meyer
Department of Chemistry
University of North Carolina at Chapel Hill
Venable and Kenan Laboratory CB 3290
Chapel Hill, North Carolina 27599-3290, USA
E-mail: tjmeyer@email.unc.edu

Prof. Herbert W. Roesky
Institut für Anorganische Chemie
der Universität Göttingen
Tammannstraße 4
D-37077 Göttingen, Germany
E-mail: hroesky@gwdg.de

Prof. Arndt Simon
Max-Planck-Institut für
Festkörperforschung
Heisenbergstraße 1
70569 Stuttgart, Germany
E-mail: simon@simpow.mpi-stuttgart.mpg.de

Structure and Bonding
Now Also Available Electronically

For all customers with a standing order for Structure and Bonding we offer the electronic form via LINK free of charge. Please contact your librarian who can receive a password for free access to the full articles by registration at:

http://link.springer.de/orders/index.htm

If you do not have a standing order you can nevertheless browse through the table of contents of the volumes and the abstracts of each article at:

http://link.springer.de/series/sb/
http://link.springer-ny.com/series/sb/

There you will also find information about the

- Editorial Board
- Aims and Scope
- Instructions for Authors

Foreword

It is a great personal pleasure to be introducing the 100th Volume of *Structure and Bonding*. The series was launched when I was an undergraduate and it proved to be an important resource for my initial exploration of inorganic chemistry and subsequently as a medium for the publication of some of my reviews when I developed a research interest in valence problems. Therefore, the series has served as a historical backdrop for my association with the subject. Initially the series was set up with the following intention: "A valuable service is performed by bringing together up-to-date authoritative reviews from the different fields of modern inorganic chemistry, chemical physics and biochemistry, where the general subject of chemical bonding involves a metal and a small number of associated atoms.... We are particularly interested in the role of the "complex metal-ligand" moiety and wish to direct attention towards borderline subjects.... We hope that this series may help bridge the gaps between some of these different fields and perhaps provide in the process some stimulation and scientific profit to the reader".

Of course in the last 30 years the subject has progressed and the initial reviews on ligand field theory, hard and soft acid base phenomena and bonding issues in cluster compounds have been progressively replaced by reviews which summarise and interpret interdisciplinary research at the borders of chemistry, biology and physics and reflect the breaking down of the barriers between inorganic, organic and physical chemistry. Therefore, in 1998 the Editorial Board refined the aims of the series in the following terms: "We expect the scope of *Structure and Bonding* series to span the entire periodic table and address structure and bonding issues wherever they may be relevant. Therefore, it is anticipated that there will reviews dealing not only with the traditional areas of chemical bonding based on valence problems and dynamics, but also nano-structures, molecular electronics, supramolecular structure, surfaces and clusters. These represent new and emerging areas of chemistry".

With these aims in mind it is noteworthy that the current volume effectively reinforces and illustrates these ideals and is titled π-*Electron Magnetism from Molecules to Magnetic Materials*. Professor Veciana is to be congratulated on bringing together a series of reviews on organic molecular magnets. This area has considerable potential applications, but also presents considerable challenges for the synthetic, solid state and theoretical chemists. An underlying theme is how systematic variations in the molecular structure of unsaturated organic molecules and the packing modes of the molecules may control their collective solid state magnetic properties. It is interesting how magneto-chemistry which historically was an area associated with transition metal ions in 1950's has evolved into a topic of current interest to organic chemists. An understanding of the relationships between the packing modes of the molecules and the intermolecular magnetic interactions provides an equivalent challenge to the interpretation of the magnetic properties of transition metal complexes using crystal field theory half a century ago.

May I take this opportunity of thanking Springer-Verlag and their editorial staff for all those behind the scenes activities which have enabled us to make this 100th volume a celebration and the members of the Editorial Board who have had the imagination to regularly propose new and exciting topics for individual volumes. It takes considerable enthusiasm to approach and badger colleagues to produce the manuscripts reasonably close to the submission dates. They may have lost friends in the process, but the scientific community generally has gained from the presence on library shelves of such a fine series.

Oxford, February 2001   Michael Mingos

Preface

The notion of *organic molecular materials* with magnetic properties started three decades ago as a mere dream of some members of the chemical community. The goal was to create a collection of molecules, composed only of light elements (C, H, N, O, S), with unpaired electrons located in orbitals with a π symmetry, in which the whole material possesses the macroscopic magnetic properties exhibited by some classical metals or oxides – ferromagnetism, ferrimagnetism, etc. Such magnetic properties are unattainable from the isolated molecular building-blocks and took a couple of decades to be realized after its first proposal because of the subtle and complex structural and electronic interactions required by these materials.

There is an awareness among scientists and technologists that *organic magnetic materials* offer considerable potential for applications in many diverse areas owing to the intrinsic characteristics of organic molecules (processability, transparency, flexibility, etc.). The skills of the organic chemist in allowing subtle variation in molecular structure of building blocks allow solid state characteristics to be fine-tuned. It is this rich potential for extreme diversity that gives *organic molecular magnetic solids* an important advantage. An underlying theme is, therefore, in the systematic variation of molecular structure and its effect on molecular packing and hence in the intermolecular electronic interactions that finally control the solid state magnetic property. An understanding of the relationship existing between packing arrangements in the solid and the intermolecular magnetic interactions are essential and many of the present developments in this area are linked to this aspect. This book contains a series of contributions chosen to highlight developments and methodologies in this direction of research. It is not intended to cover either all the topics of research in organic magnetic materials or all details of each of these topics. Thus, high-spin macromolecules have been deliberately excluded from this volume because they probably would require an entire volume of this book series.

All topics contained in this volume deal with organic molecule-based magnetic materials in which the unpaired spins of the building-block units are located on π-orbitals conferring special characteristics that differentiate them from materials based on transition metal complexes. The first two chapters discuss the most representative examples of purely organic 3-D magnets as well as the through-space ferromagnetic interactions responsible of the bulk magnetic orderings. The third chapter shows the use of free radicals as ligands of transition metals for the design and preparation of magnets showing high critical temperatures while the fourth chapter is devoted to the most promising approach existing nowadays for increasing the critical temperature of molecular magnets: the weak (canted) ferromagnetism. The goal of the next chapter is to review the magnetic properties of the different materials based on C_{60} since this building block may be considered as a metal-like unit permitting the obtaining of materials with metallic, superconducting and ferromagnetic properties. The last chapter describes a new kind of exotic magnetic materials – spin ladder – which results from the application of general concepts of supramolecular chemistry and shows the potentiality of *organic magnetic materials*.

I hope that this volume will be of use to researchers working in this area and also serve as a source of reference and an inspiration for others working in its frontiers in chemistry, physics, or materials science.

Cerdanyola, May 2001 \hfill Jaume Veciana

Contents

An Organic Radical Crystal Showing Spontaneous
Ferromagnetic Order................................... 1
M. Kinoshita

The Mechanism of the Through-Space Magnetic Interactions
in Purely Organic Molecular Magnets 33
J.J. Novoa, M. Deumal

Metal-Aminoxyl-Based Molecular Magnets................... 61
K. Inoue

Magnetic Properties of Thiazyl Radicals.................... 93
J.M. Rawson, F. Palacio

Magnetism in Fullerene Derivatives........................ 129
D. Arčon, K. Prassides

Molecular Compounds Showing a Spin Ladder Behaviour......... 163
C. Rovira

Author Index Volumes 1–100 189

Contents of Volume 98

Localized to Itinerant Electronic Transition in Perovskite Oxides

Volume Editor: J.B. Goodenough

General Considerations
J.B. Goodenough

Transport Properties
J.B. Goodenough, J.-S. Zhou

Local Atomic Structure of CMR Manganites and Related Oxides
T. Egami

Optical Spectroscopic Studies of Metal-Insulator Transitions in Perovskite-Related Oxides
S.L. Cooper

An Organic Radical Crystal Showing Spontaneous Ferromagnetic Order

Minoru Kinoshita

Department of Materials Science and Engineering, Science University of Tokyo in Yamaguchi, Onoda-shi, Yamaguchi 756-0884, Japan
E-mail: mkino@ed.yama.sut.ac.jp

An example of purely organic ferromagnets was first found nine years ago in the β-phase crystal of the p-nitrophenyl nitronyl nitroxide radical (abbreviated as p-NPNN), which consists only of light elements, H, C, N, and O. This finding was prompted by extensive studies on the mechanism governing intermolecular ferromagnetic interaction in the galvinoxyl radical crystal. In the mid-1980s, only a few organic radicals were known to exhibit intermolecular ferromagnetic interaction. Galvinoxyl had been a typical example of such radicals. Strategies learned from galvinoxyl were first applied to p-NPNN, and the ferromagnetic transition was observed at 0.6 K. This article describes the progress of the research in this regards and the recent developments since then. The antiferromagnetism of the γ-phase crystal of p-NPNN is also briefly described.

Keywords: Organic ferromagnet, Intermolecular ferromagnetic interaction, Nitronyl nitroxide radicals, Spin polarization, p-NPNN

1	Introduction	2
2	Magnetism and Chemical Bonding	3
3	Galvinoxyl	4
3.1	Magnetic Properties of the Neat Crystal	4
3.2	Magnetic Properties of Mixed Crystals	5
3.3	Mechanism of Ferromagnetic Interaction	8
3.3.1	SOMO-SOMO Overlap	8
3.3.2	Charge-Transfer Interaction	9
4	Conditions for Ferromagnetic Interaction in Organic Crystals .	12
5	Ferromagnetism of p-Nitrophenyl Nitronyl Nitroxide	14
5.1	Electronic Structure of p-NPNN	14
5.2	Crystal Structure of p-NPNN	15
5.3	Magnetic Properties of p-NPNN Crystal	15
5.4	Transition to a Ferromagnetic Ordered State	17
5.5	Other Evidence of Ferromagnetism	19
5.5.1	Field Effect on Heat Capacity	19
5.5.2	Zero-Field Muon Spin Rotation	20

5.6	Pressure Induced Ferro- to Antiferromagnetic Transition	22
5.7	Charge-Transfer Mechanism	24
5.8	Lattice Constants Under High Pressure	25
6	Antiferromagnetism of γ-Phase Crystal of p-NPNN	27
7	Other Ferromagnetic Organic Crystals	30
8	References	30

List of Abbreviations

a.c.	alternating current
AFM	antiferromagnetic
CT	charge-transfer
FM	ferromagnetic
FOMO	fully occupied molecular orbital
galvinoxyl	4-[[3,5-bis(1,1-dimetylethyl)-4-oxo-2,5-cyclohexadien-1-ylidene]methyl]-2,6-bis(l,1-dimethylethyl)phenoxy
hydrogalvinoxyl	4-[[3,5-bis(1,1-dimethylethyl)-4-hydroxyphenyl]methylene]-2,6-bis(l,1-dimethylethyl)-2,5-cyclohexadien-1-one
INDO	incomplete neglect of differential overlap
M	magnetization
MO	molecular orbital
NHOMO	next highest occupied molecular orbital
NLUMO	next lowest unoccupied molecular orbital
p-NPNN	p-nitrophenyl nitronyl nitroxide (2-(4-nitrophenyl)-4,4,5,5-tetramethyl-4,5-dihydro-1H-imidazol-1-oxyl-3N-oxide)
SOMO	singly occupied molecular orbital
T_C	critical temperature
UHF	unrestricted Hartree-Fock
ZF-μSR	zero-field muon spin rotation
χ	magnetic susceptibility

1
Introduction

The study of magnetic properties of organic radical crystals has attracted much attention in the last few years. In particular, a movement to search out bulk ferromagnetism in organic materials composing exclusively of light elements was quite active in the 1980s. After the achievement of organic bulk ferromagnetism in 1991 [1], the magnetic properties of a variety of stable organic radicals such as phenoxy, nitroxide, and verdazyl radicals have been extensively studied. Among them, the derivatives of nitronyl nitroxide are very stable and suited for the study of their solid state properties. The prominent feature of organic radicals is twofold. One is the highly isotropic nature of electron spins, which results from weak spin-orbit coupling in a molecule composed only of light elements such as H, C, N, and O. The other is the distribution of spin densities over a molecule of π-electron system. These result in little anisotropy of the magnetic interactions. Therefore, the magnetism of organic crystals is, in most cases, analyzed on the basis of ideal Heisenberg spins. The distribution of spin densities also causes a complicated interaction scheme between adjacent radicals in a crystal, and gives rise to a variety of magnetic behavior. This is another feature of magnetism of organic crystals in comparison to that of inorganic crystals, in which localized character of electron spins is dominant.

2
Magnetism and Chemical Bonding

Magnetism and the concept of chemical bonding are closely related to each other apart from the energy scale involved therein. In chemistry, a single bar indicates a single covalent bond, which corresponds to a pair of electrons coupled antiferromagnetically, that is with their magnetic moments in antiparallel fashion in accordance with the Pauli exclusion principle. In other words, a pair of electrons gets together in a small space between the atoms and becomes an adhesive agent for positively charged nuclei. Such a bond is quite stable in most cases, although these electrons repel each other in a small space. The stability comes from the overlap (or non-orthogonality) of the atomic orbitals associated with the bonded atoms. In the Heitler–London model for a hydrogen molecule (thus in terms of valence bond theory), it is the finite overlap integral which brings the stability to the spin singlet ground state.

A similar situation occurs when two organic radicals are brought together closely, resulting in bond formation. An example is the triarylmethyl radical in solution, which yields hexaarylethane. In this case, a true covalent bond is formed between the methyl carbon atoms and the unpaired electrons are strongly coupled antiferromagnetically. From these and other examples, bond formation can be said to be equivalent to antiferromagnetic (AFM) spin coupling.

There is the possibility of another kind of bond formation in the case of organic radicals. Some π-radicals, such as diphenylpicrylhydrazyl, nitroxides, phenoxyl derivatives, and many ion radical salts, are very stable even in the crystalline state. In these crystals, weak AFM couplings are usually recognized by magnetic measurements (i.e., the magnetic susceptibilities tend to become smaller at low temperature than those expected from the Curie law). From the analogy to the equivalence of chemical bond and AFM spin coupling, a weak chemical bond can be said to form between adjacent radicals in the crystal, where any atom in the radical species may not be specified to be involved in the bond formation. The bond is formed between radical molecules themselves, i.e., the radical molecule as a whole behaves like an atom. These situations are illustrated in Fig. 1.

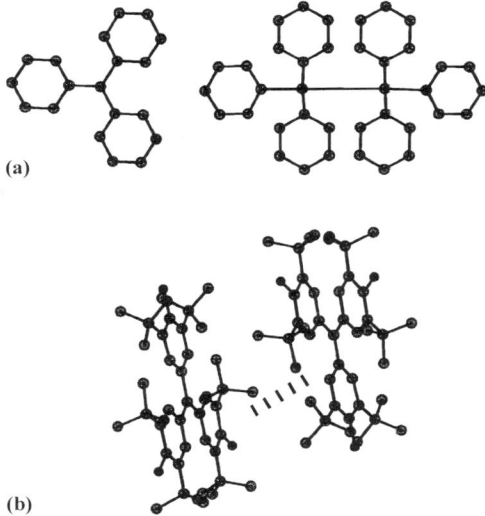

Fig. 1a, b. Schematic illustration of bond formation between radicals: **a** real bond formation; **b** weak intermolecular bond formation

On the other hand, when the overlap between the orbitals of the unpaired electrons is small or zero, the triplet state or the state of parallel spin alignment is stable. This is the case with an oxygen molecule or carbene type radical. The physical basis of the spin parallel arrangement, in these molecules, is the Hund rule (or orthogonality). In the case of ferromagnetism, the spin magnetic moments on radical species align in a parallel fashion throughout the magnetic domain. Then, from the chemical point of view, no bond forms between adjacent radicals in the crystal. Therefore, to design an organic radical ferromagnet is, in a sense, a kind of chemistry that does not make a chemical bond between unpaired electrons on adjacent radicals.

3
Galvinoxyl

3.1
Magnetic Properties of the Neat Crystal

Until the mid-1980s, only two organic radicals were known to exhibit intermolecular ferromagnetic (FM) interaction. One of them is the galvinoxyl (4-[[3,5-bis(1,1-dimetylethyl)-4-oxo-2,5-cyclohexadien-1-ylidene]methyl]-2,6-bis(1,1-dimethylethyl)phenoxy) radical, in which the ferromagnetic interaction was first observed by Mukai in 1969 [2]. In the inset of Fig. 2, the temperature dependence of the paramagnetic susceptibility of the galvinoxyl crystal is shown [3]. The susceptibility above 85 K follows the Curie-Weiss law with a ferromagnetic Weiss constant of 19 K. The crystal, however, undergoes

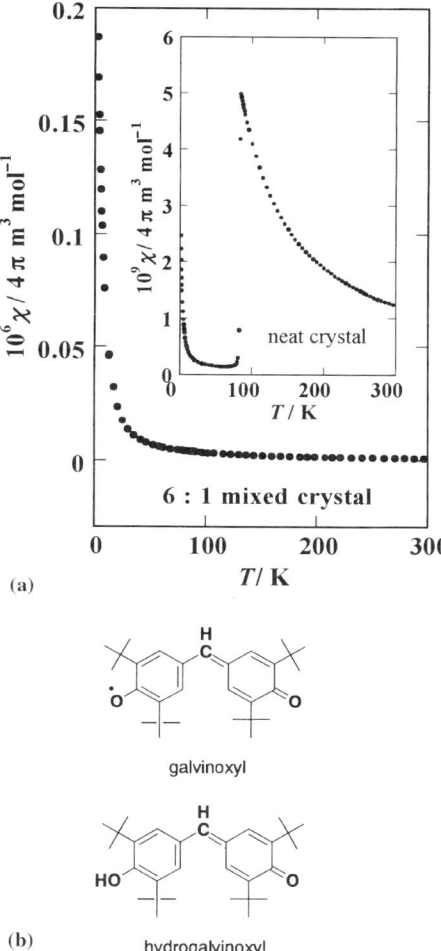

Fig. 2a, b. Paramagnetic susceptibilities of the neat galvinoxyl crystal (*inset*) and of 6:1 mixed crystal of galvinoxyl and hydrogalvinoxyl (*main frame*)

a phase transition at 85 K and becomes almost diamagnetic below 85 K. This had caused difficulty in studying FM interaction in the temperature region of the Weiss constant.

3.2
Magnetic Properties of Mixed Crystals

It was found that the phase transition is suppressed in making mixed crystals with a small amount of hydrogalvinoxyl (4-[[3,5-bis(1,1-dimethylethyl)-4-hydroxyphenyl]methylene]2,6-bis(1,1-dimethylethyl)-2,5-cyclohexadien-1-one)

[3–5]. Hydrogalvinoxyl is a precursory closed shell compound for which the crystal structure is known to be isomorphous to that of galvinoxyl, but phase change does not occur [6]. The temperature dependence of the susceptibility of the 6:1 mixed crystal is shown in the main frame of Fig. 2. As is shown in Fig. 3, the reciprocal susceptibility of the mixed crystal crosses the temperature axis in the positive region, thereby confirming that the FM interaction is maintained even in the mixed crystal down to the cryogenic temperature.

To comprehend the above matter in detail, the magnetic field dependence of the magnetization at 2 K was measured by making three types of the crystals of 4:1, 6:1, and 9:1 mixing ratios [4]. The result is shown in Fig. 4. Depicted with the dotted curves are the theoretical curves of the Brillouin function for the

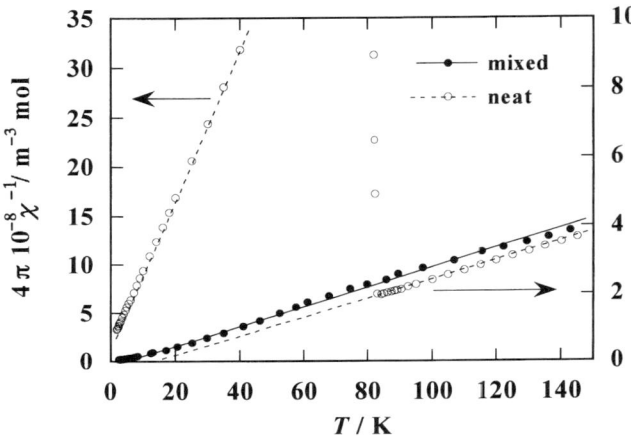

Fig. 3. Reciprocal susceptibilities of the neat galvinoxyl crystal and 6:1 mixed crystal

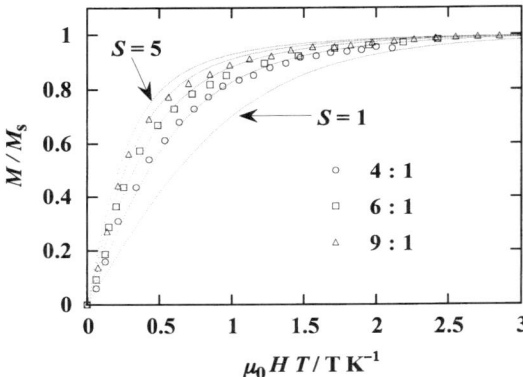

Fig. 4. Magnetization isotherms of the 4:1, 6:1 and 9:1 mixed crystals. The *dotted curves* represent the theoretical ones for $S = 1, 2, 3, 4$, and 5

spin quantum numbers $S = 1, 2, 3, 4$, and 5. When compared with the theoretical curves, it is seen that the experimental result for the crystals of n:1 mixing ratio almost corresponds to the theoretical curve for $S = n/2$. Therefore it has turned out that n pieces of the galvinoxyl radicals of $S = 1/2$ are amassed as an average, and their magnetic moments are aligned in parallel at low temperatures.

As shown in Fig. 5, the crystal of galvinoxyl belongs to the monoclinic system ($C2/c$) and the almost planar radical molecules are arranged with the plane facing with each other along the c-axis at an equal interval [7]. If it is assumed that the one-dimensional interaction along the c-axis is effective, the above results are easy to understand. That is, the one-dimensional chain of radicals is divided into a number of segments comprised, on average, of n radicals separated by the closed shell molecules having similar molecular structure, and the magnetic moments of the radicals are aligned in parallel within the individual segments [5].

The magnitude of the exchange interaction was obtained by using paramagnetic resonance absorption [8]. In this case, we used a dilute mixed crystal containing about 4% of the radical and observed the temperature dependence of the absorption intensity of the $\Delta m = \pm 2$ transitions, which are typical of the radical pairs existing statistically. The result is shown in Fig. 6. The ordinate, the product of the intensity and temperature, is the quantity proportional to the occupation number in a magnetic state (the triplet state in this case). The occupation number increases as the temperature decreases. Evidently the ground state is a magnetic state. The energy difference between the magnetic and non-magnetic states can be obtained as 1.5 meV by fitting to the theoretical equation for the susceptibility of radical pair. The value

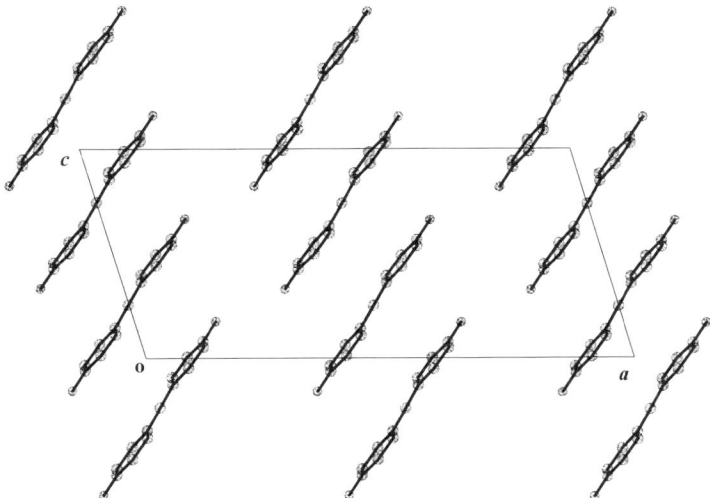

Fig. 5. Crystal structure of galvinoxyl. The *tert*-butyl groups are omitted

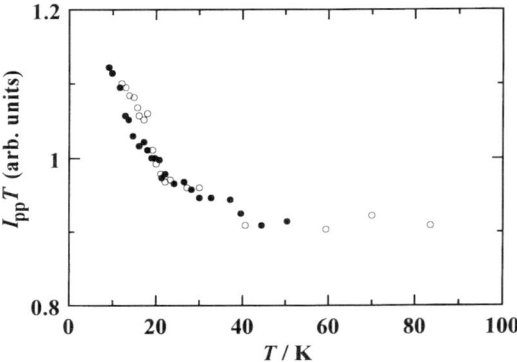

Fig. 6. The temperature dependence of paramagnetic absorption intensity of the $\Delta m = \pm 2$ transitions observed in the dilute (4%) mixed crystal of galvinoxyl in hydrogalvinoxyl

corresponds well to the Weiss constant obtained from the susceptibility measurements.

3.3
Mechanism of Ferromagnetic Interaction

3.3.1
SOMO-SOMO Overlap

The mechanism governing the FM interaction is, of course, related to the electronic structure of the galvinoxyl radical and its crystal. We first examined the molecular orbitals (MO) of galvinoxyl by the INDO UHF method.

As was mentioned above, the AFM interaction most frequently encountered among a number of organic radicals is explained by means of an analogy with the hydrogen molecules because two neighboring radicals have a tendency to form a weak covalent bond by making a pair of the unpaired electrons. For the formation of a covalent bond, overlap of the orbitals of the unpaired electrons plays an important role. With galvinoxyl where directions of the spins are prone to be in parallel, how is the overlap of the orbitals? This point was examined first.

The intermolecular overlap integral was calculated with the singly occupied molecular orbitals (SOMO) using the molecular location of the crystal structure. The value of the SOMO-SOMO overlap became very small. When calculation was made by changing the relative location of the molecules step by step, it turned out that the overlap is close to the minimum when the molecular location is in the crystal structure.

Furthermore, when the calculation process was examined it was found that relatively large positive and negative contributions cancel out each other, resulting in a small value. Since the π-orbitals are spread out over the molecule, the partial overlaps become positive and negative here and there and cancel out each other. As a result, the SOMOs on the adjacent radicals are nearly orthogonal and degenerate. This is a situation quite similar to the case of the d-orbitals of an isolated transition metal ion. In the case of d-orbitals on one center, the partial overlap becomes positive and negative, but cancels out to null and the d-orbitals are orthogonal and degenerate. The exchange integral becomes positive in such a case and the high spin state is stabilized in accordance with the Hund rule. With galvinoxyl, even though it is a multi-center molecule and the intermolecular interaction is now considered, the Hund rule may be applied in a modified manner and the spin state must be stable when the spins are maintained in parallel. This is the so-called potential exchange.

3.3.2
Charge-Transfer Interaction

The average distance between the neighboring galvinoxyl radicals along the c-axis is as much as 4.05 Å, and the positive and negative overlaps themselves are not very large. Accordingly it appears that the interaction large enough to be 1.5 meV cannot be explained only by the static exchange interaction mentioned above. Some other mechanism, namely charge-transfer interaction, should also be taken into account.

The π-MO energy levels of galvinoxyl near the SOMO level are depicted in Fig. 7a [9–11], where the α-spin orbitals and the corresponding β-spin orbitals are connected with dotted lines. The most conspicuous feature of this figure is that there exists the next highest occupied MO level of β-spin (NHOMO-β)

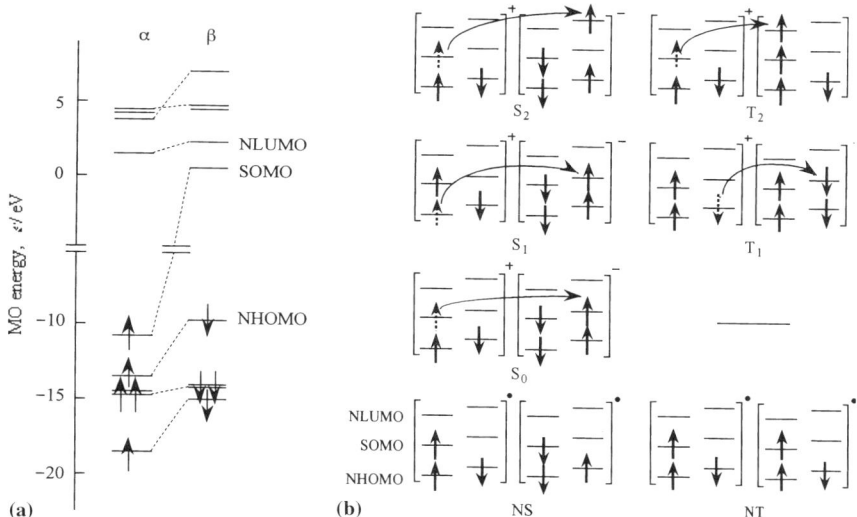

Fig. 7. a Molecular orbital energy levels of galvinoxyl calculated by the INDO UHF method. Only the levels near SOMO are shown. **b** The electronic configurations in the radical pair coupled by CT interaction

situated higher than the SOMO of α-spin (SOMO-α). The exchange interaction within the molecule is great enough to stabilize the SOMO-α. In other words, the spin correlation causes a large spin polarization effect in galvinoxyl.

In Fig. 7b, some low energy charge-transfer (CT) configurations are shown for the neighboring pair of radicals. NS and NT are the no-bond structures of singlet and triplet multiplicity and S_i and T_i (i = 0, 1, 2) are the excited CT state configurations, respectively. Among the excited CT configurations, S_0 is the lowest and the resonance between S_0 and NS usually stabilizes NS, resulting in an AFM interaction. This is the reason why most organic radicals exhibit AFM intermolecular coupling. With galvinoxyl, however, the SOMO-SOMO overlap is quite small as mentioned above. Therefore, the stabilization of NS by admixture of S_0 is expected to be minimized. On the other hand, T_1 and T_2 must be lower in energy than S_1 and S_2, respectively. In particular, T_1 and T_2 are much stabilized with respect to S_1 and S_2 in galvinoxyl because of the large spin polarization effect. Thus stabilization of NT by admixture of T_1 and T_2 is expected to overweigh that of NS and the ground state becomes magnetic. It is in this way that the FM interaction is brought about in galvinoxyl.

For the sake of assurance, the intermolecular overlap integrals, which are thought to be proportional to transfer integral, are calculated for the orbitals relevant to the above configurations with the molecular arrangement in the crystal. The results are given in Table 1 [9–11]. As is seen, the overlap integrals for the T_1 and T_2 configurations are, respectively, larger than those for S_1 and S_2. Thus it is concluded that the off-diagonal interaction is also favorable for the NT stabilization.

From these observations, it is concluded that the cooperative effect of the spin polarization caused by a large exchange interaction within a molecule and the charge transfer interaction among molecules is essential for the FM interaction of the galvinoxyl crystal and probably of organic radical crystals in general [9]. The arguments described in this section may be summarized schematically as those shown in Fig. 8, where the states, S_2 and T_2, are omitted for the sake of simplicity. In the case of FM coupling, the electrons in SOMO and NHOMO play an important role to make intermolecular bonds (Fig. 8a) because the overlap between them is large enough. In other words, the molecular design giving an opportunity of participating in a bond formation to the electrons in NHOMO, or more generally those in FOMOs (fully occupied MOs) is of crucial importance for FM interaction. The case of AFM interaction is shown in Fig. 8b, where the overlap between SOMOs are large enough to make a weak intermolecular covalent bond.

Table 1. The calculated overlap integrals for the various CT configurations

S_0	$\langle$SOMO-α|SOMO-$\beta\rangle$	0.72×10^{-3}
T_1	$\langle$NHOMO-β|SOMO-$\beta\rangle$	1.60×10^{-3}
S_1	$\langle$NHOMO-α|SOMO-$\beta\rangle$	0.87×10^{-3}
T_2	$\langle$SOMO-α|NLUMO-$\alpha\rangle$	2.73×10^{-3}
S_2	$\langle$SOMO-α|NLUMO-$\beta\rangle$	1.33×10^{-3}

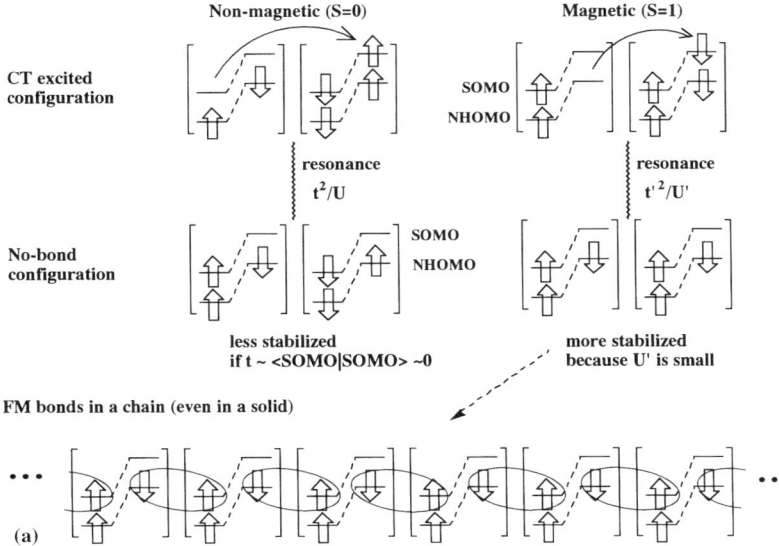

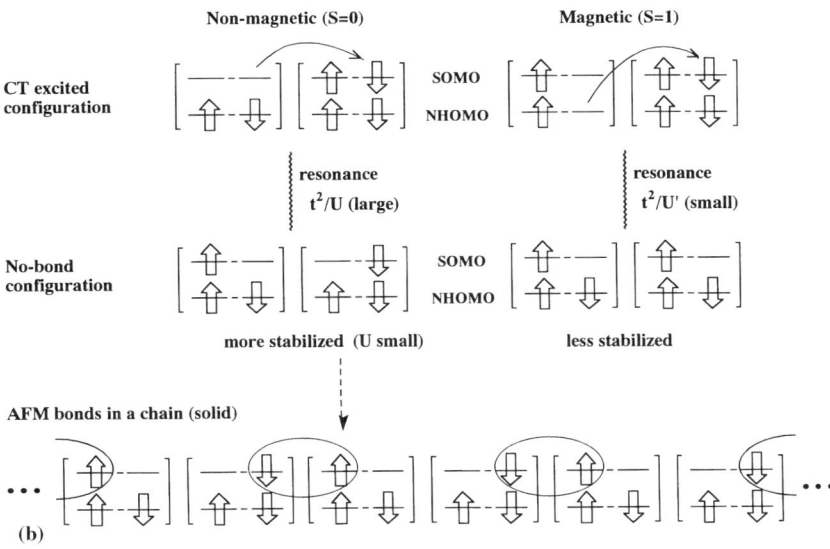

Fig. 8a, b. Schematic illustrations of; **a** ferromagnetic; **b** antiferromagnetic intermolecular bond formation. The bonds are shown by the *elliptic circles*. Note that the spins on NHOMO are aligned parallel in FM configuration, because the bonds are formed between the electrons in SOMO and NHOMO, while the intermolecular bonds are formed between the electrons only in SOMOs in the case of AFM configuration

4
Conditions for Ferromagnetic Interaction in Organic Crystals

In the preceding section we have examined the FM interaction in the galvinoxyl crystal. Now we extract the conditions for FM interaction applicable to other organic radicals from the study on galvinoxyl.

As the property of radicals, it is first of all necessary for the spin polarization to be large enough. In other words, the exchange interaction within a molecule should be large enough. In order to obtain large exchange interactions in a molecule consisting of a number of atoms, it is recommended to use radicals having heteroatoms, such as nitrogen and oxygen. The unpaired electron resides largely on such electronegative heteroatoms, and interacts with the nonbonding electrons in them. This interaction is expected to be very large because of its one-center character and virtually determines the magnitude of the exchange interaction of the molecule, even though the unpaired electron is delocalized over the molecule. In galvinoxyl, the oxygen atoms at both terminals play this role. When the exchange interaction within a molecule is very large, the CT configuration T_1 of Fig. 7b is greatly stabilized and the energy of T_1 is close to that of S_0. This means that the exchange interaction almost overwhelms the energy interval between SOMO and NHOMO; namely SOMO and NHOMO behave as if they were degenerate. In order for this to be realized, a molecule having an extended π-electron system is preferred because the interval between SOMO and NHOMO is related to the size of π-electron system. The extension of this argument would lead us to the situation where T_1 is below S_0 energetically; this is exactly the situation proposed by McConnell for FM coupling in 1967 [12].

Next is the problem of crystal structure. As mentioned above, the SOMO-SOMO overlap should be minimized. This presents a problem of where to locate the radicals in a crystal. Although it is not easy to control the crystal structure, it has been attempted by changing the position or size of chemical substituents and by introducing a charge or hydrogen bonding. However, the condition favoring ferromagnetism is herein pursued from the viewpoint of the electronic property of radicals.

The galvinoxyl radical is known as a radical with large spin polarization in the field of chemistry, and has been studied as an example of a radical which possesses the so-called negative spin density. The negative spin density is usually observed in a radical of an odd-alternate π-electron system. The feature of an odd-alternate system lies in SOMO; SOMO has nodes alternately on the bonded atoms and lobes with opposite polarities on each of the neighboring atoms. Therefore, when two planar odd-alternate radicals are located close to each other in a face-to-face manner and are relatively deviated by one bond or an odd number of bonds, the local overlap is expected to become small through effacing with each other. In Fig. 9, a schematic diagram of SOMO of galvinoxyl and the relative location of the neighboring radicals are illustrated. It is easily seen that the local overlap is

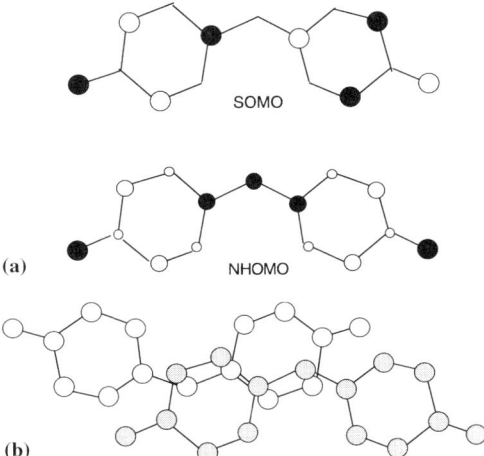

Fig. 9. a Schematic drawing of SOMO and NHOMO of galvinoxyl (the *tert*-butyl groups are omitted). *Filled and open circles* show the difference in polarity of the *p*-orbitals. **b** The arrangement of adjacent molecules viewed along the normal to the best-fit molecular plane

well effaced, resulting in a minimal SOMC-SOMO overlap throughout the whole of the molecules. Thus there is a high possibility of achieving such relative positioning of small SOMO-SOMO overlap by the use of radical of the odd-alternate system [9].

This situation may be evaluated from another point of view. Even though the charge density is zero, a finite spin density is induced at the nodes by the electronic correlation (or exchange interaction) effect. The induced spin density is negative in the sense that the direction of the induced spin is opposite to that of the unpaired electron. Since there are positive spin densities on the other atoms located at the lobes, the relative face-to-face location of two radicals deviating by one or an odd number of bonds corresponds to the situation where the atom of negative spin density in one molecule overlaps the atom of positive spin density in another molecule. This situation was also proposed for FM interaction by McConnell in 1963 [13].

Another advantage of the use of the odd-alternate system is that there is a high probability of the overlap of SOMO with NHOMO or NLUMO being enhanced once such a relative location is established, because the charge density is usually distributed over the molecule in the π-MOs other than SOMO.

The arguments given in this section are summarized as follows. The radicals of the odd-alternate system having heteroatoms and relatively developed π-conjugation provide a high possibility of realizing the intermolecular FM interaction in a crystal [9].

5
Ferromagnetism of p-Nitrophenyl Nitronyl Nitroxide

5.1
Electronic Structure of p-NPNN

The first radical employed fulfilling the conditions in the preceding section is p-nitrophenyl nitronyl nitroxide (2-(4-nitrophenyl)-4,4,5,5-tetramethyl-4,5-dihydro-1H-imidazol-1-oxyl-3N-oxide, abbreviated as p-NPNN hereafter). Its large spin polarization effect has long been an object of research in the field of chemistry. When the molecular orbitals are actually calculated, the spin polarization effect manifests itself in a more obvious manner than in galvinoxyl; i.e., the fourth occupied MO for the β-spin maintains higher energy than SOMO-α, as shown in Fig. 10.

Also, the charge density in SOMO is mostly concentrated on the two NO moieties and only a little is distributed over the other parts of the molecule. Accordingly, it is expected that the SOMO-SOMO overlap is minimized provided that the NO moieties of the neighboring radicals do not approach each other in the crystal (this actually holds in the β-phase crystal, see below). On the other hand, as the charge distribution in other orbitals ranges over the whole molecule, the overlap with SOMO becomes large. These are quite favorable for FM interaction, as discussed above.

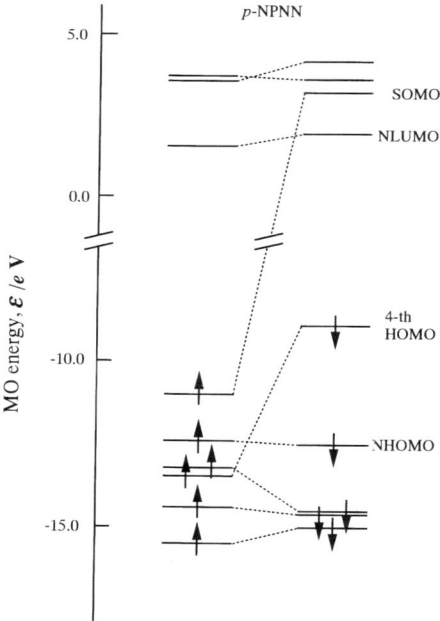

Fig. 10. Molecular orbital energy levels of p-NPNN calculated by the INDO LJHF method, showing large spin polarization effect. Only the levels near SOMO are shown

5.2
Crystal Structure of p-NPNN

There are four polymorphic forms, α-, β-, γ-, and δ-phases, known in p-NPNN [14–17]. Each of them can be separately prepared by properly adjusting the conditions for depositing crystals from solutions [18]. The crystallographic constants of these phases are given in Table 2 and the structure of the β-phase is shown in Fig. 11. The orthorhombic β-phase is the most stable form, and the other forms are subject to change to the β-phase when they are kept at room temperature or below room temperature. Hereafter, details of the magnetism of the β-phase are described.

5.3
Magnetic Properties of p-NPNN Crystal

The magnetic properties of the β-phase crystal of p-NPNN was first reported in 1989 [19]. The temperature dependence of the paramagnetic susceptibility follows the Curie–Weiss law and gives a ferromagnetic Weiss constant of about 1 K. In Fig. 12, the result of measurements with $H//a$ is shown [1, 18, 20]. There is no sizable anisotropy in the susceptibility, and nearly the same results are obtained, within the range of errors, in the other directions of the applied field.

Since the Weiss constant θ is very small, the FM interaction should also be checked by measuring the field dependence of the magnetization at low temperatures [1, 18, 20]. Figure 13 shows the results. The magnetization curves at several temperatures are unified into the single curve of the Brillouin function for $S = 1/2$ by the molecular field correction. In this case, the best fit is obtained with the coefficient of $\lambda = 2.8$ Oe mole emu^{-1} ($H_{\text{eff}} = H + \lambda M$), yielding an FM Weiss constant of $\theta = 1.1$ K in agreement with the result of susceptibility measurements. In addition, this experiment assures that the sample is not contaminated with an FM impurity.

Table 2. Crystal structures of p-nitrophenyl nitronyl nitroxide

	α-phase	β-phase		γ-phase	δ-phase
T/K	≈300	≈300	6	≈300	≈300
System	monoclinic	orthorhombic		triclinic	monoclinic
Space group	$P2_1/c$	F2dd		$P\bar{1}$	$P2_1/c$
a/Å	7.302	12.347	12.16	9.193	8.963
b/Å	7.617	19.350	19.01	12.105	23.804
c/Å	24.677	10.960	10.71	6.471	6.728
α/[(π/180) rad]				97.35	
β/[(π/180) rad]	93.62			104.44	104.25
γ/[(π/180) rad]				82.22	
Z	4	8	8	2	4
V/Å³	1369.7	2618.5	2475.7	687.6	1391.3
Density, ρ/g cm^{-3}	1.354	1.416	1.498	1.349	1.333

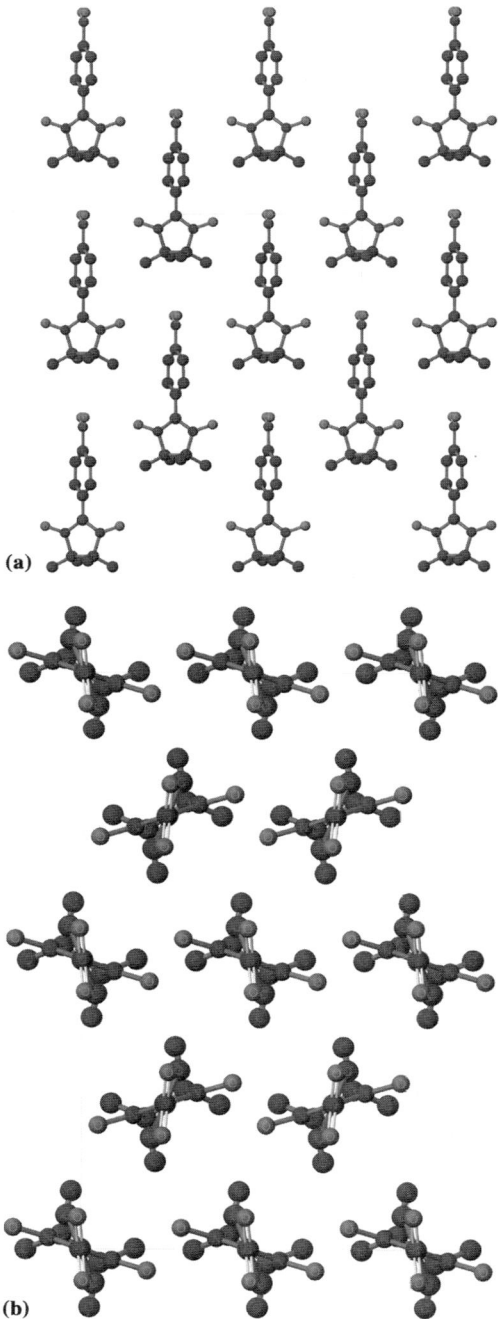

Fig. 11a, b. Structure of the orthorhombic β-phase crystal of p-NPNN: **a** molecules on the ac-plane; **b** view along the a-axis

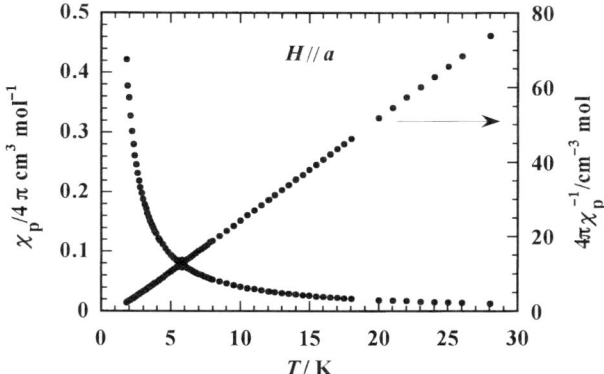

Fig. 12. The temperature dependence of paramagnetic susceptibility of the β-phase crystal of p-NPNN

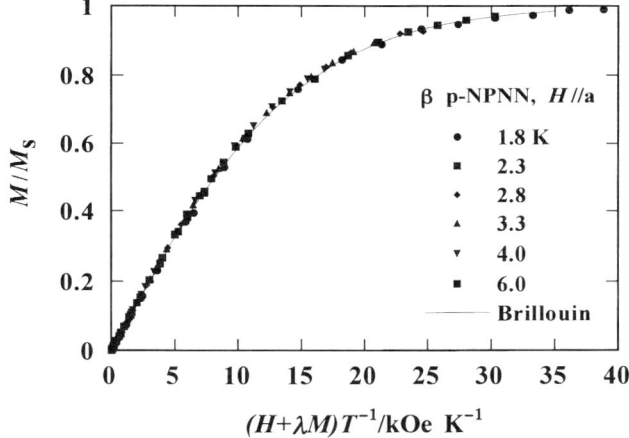

Fig. 13. The field dependence of magnetization of the β-phase crystal of p-NPNN at several temperatures. The curves are analyzed in terms of the molecular field approximation. The data almost coalesce into the curve of the Brillouin function for $S = 1/2$ with $\lambda = 2.8$ Oe mol emu^{-1}

5.4
Transition to a Ferromagnetic Ordered State

In Fig. 14, the temperature dependencies of the heat capacity and the a.c. susceptibility are shown. The heat capacity has a sharp peak at $T_C = 0.6$ K, and reveals the existence of a transition. The corresponding entropy amounts to 85% of $R\ln 2$ in the range up to 2 K. Thus it is concluded that the transition is magnetic and bulk-like in nature. Since the a.c. susceptibility (shown in the inset) diverges around T_C, the ordered state is, without a doubt, a ferromagnetic state.

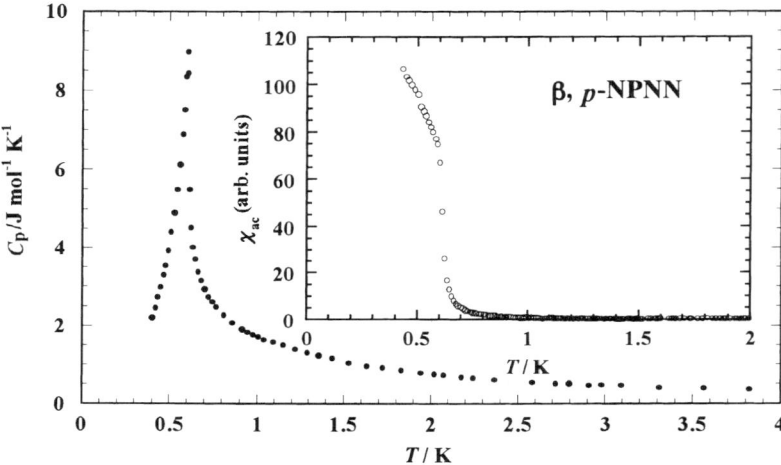

Fig. 14. The temperature dependence of heat capacity (*main frame*) and a.c. susceptibility (*inset*) of the β-phase crystal of p-NPNN

The magnetization curves measured at temperatures above and below the transition point are illustrated in Fig. 15 [1, 18, 20]. Although the curves have slight gradients at 1.22 K and 0.81 K in the paramagnetic region, the curve at 0.44 K clearly traces a hysteretic loop characteristic of ferromagnetism. The magnetization is almost saturated at about 50 Oe, and the coercive force is small. The reason for small coercive force is the small anisotropy in the g-factor and the dipole interaction. The g-factors observed in the paramagnetic

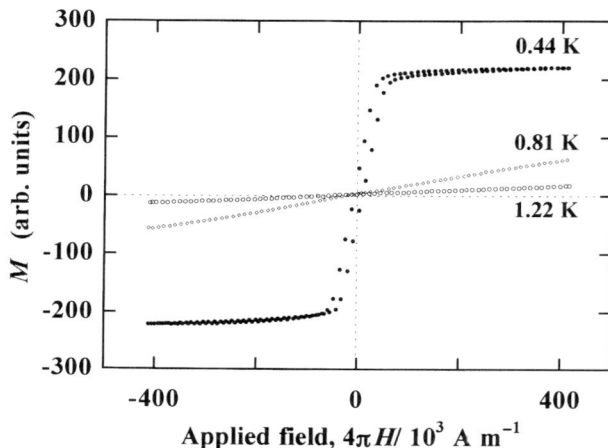

Fig. 15. The field dependence of magnetization of the β-phase crystal of p-NPNN above and below the transition temperature (0.6 K). The field has been corrected for the demagnetization

resonance are $g_a = 2.0070$, $g_b = 2.0030$, and $g_c = 2.0106$ [18]. The linewidth also shows an almost isotropic angular dependence at room temperature.

5.5
Other Evidence of Ferromagnetism

5.5.1
Field Effect on Heat Capacity

Illustrated in Fig. 16 are the temperature dependencies of the heat capacity at various magnetic field strengths [18]. The sharp peak in the zero field is eventually slightly rounded and is shifted to the higher-temperature side as the magnetic field is increased. This is a feature of ferromagnetic materials. In the low magnetic field region, the demagnetizing field compensates the applied field up to a certain temperature, the substance behaves as if it were in the zero applied field, and the sharp peak remains as shown in Fig. 16a.

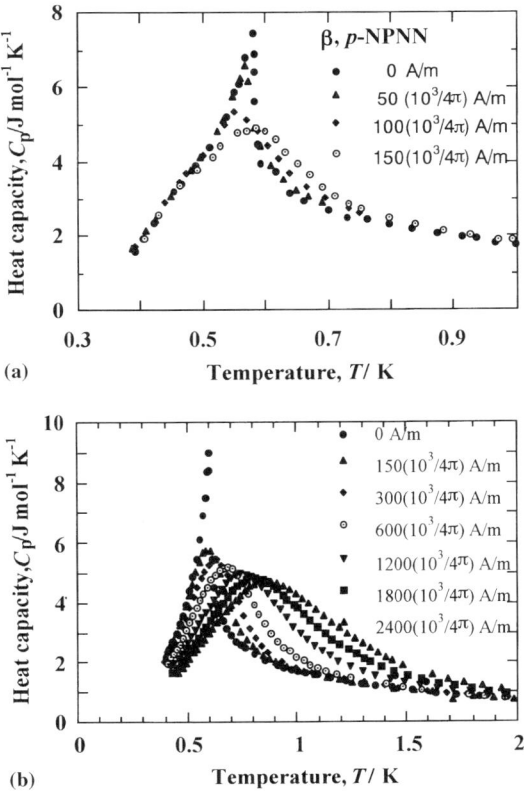

Fig. 16a, b. The temperature dependence of the heat capacity of the β-phase crystal of p-NPNN in: **a** the low; **b** the high magnetic field region

As the temperature increases above a certain temperature, however, the demagnetizing field becomes small and no longer cancels out the applied field. For FM substances, the critical temperature cannot be defined in a finite magnetic field. When there is an FM interaction among the spins, they have a tendency to align themselves in parallel along the magnetic field at very low temperature, and the spin system is ordered by a weak field even above the FM transition temperature. Thus the sharp peak of the heat capacity shifts and becomes rounded. This is shown in Fig. 16b. In the case of antiferromagnetic order, the peak remains up to a certain field strength determined by the AFM interaction (see Fig. 24). Therefore, this experiment ensures the ferromagnetism of the β-phase crystal below 0.6 K.

5.5.2
Zero-Field Muon Spin Rotation

Another piece of evidence for ferromagnetism was obtained by the measurements of the zero-field muon spin rotation (ZF-μSR). Figure 17 shows some of the results of ZF-μSR experiments conducted by Uemura and coworkers [21, 22]. The oscillating signals observed at 640 mK and 20 mK are due to the precession of the muons implanted into the crystal. Since there is no applied field, it is obvious the precession is caused by the internal field from the spontaneous magnetization. The long-lasting oscillation indicates that the muons experience a rather homogeneous local field which requires that the FM spin network is commensurate with the crystallographic structure.

This is the result for the muon spin polarization perpendicular to the b-axis. When the muon spin is polarized along the b-axis, the amplitude of the oscillating signal becomes very small (about 20% of that of the perpendicular orientation). This suggests that the spin orientation in different domains is not aligned randomly and is most likely along the b-axis. Recent FM resonance experiments by Oshima et al. [23, 24] and neutron diffraction measurements by Schweizer's group [25] also show that the magnetic easy axis is along the b-axis.

In Fig. 18, the frequency of the oscillation, which is proportional to the spontaneous magnetization, is plotted against the temperature. The dotted curve shown is the one calculated using the random phase approximation under the assumption of a three-dimensional (3-D) isotropic Heisenberg model with a specific choice of 2 J/k_B = 470 mK for the interactions with the eight neighboring radicals. These results of the ZF-μSR experiments clearly demonstrate the appearance of spontaneous magnetic order in the β-phase crystal of p-NPNN.

Although the 2 J/k_B value of 470 mK is cited above only as an average, the exchange interaction is supposed not to be of only one kind. The crystal structure of the β-phase is schematically shown in Fig. 19, where each ball denotes the p-NPNN radical. The lattice may be divided into two face-centered orthorhombic sublattices deviating by $a/4$, $b/4$, and $c/4$ to each other, similarly to a diamond or, more precisely, to a zinc blende structure. From the crystal structure, we expect that at least two kinds of interactions dominate the

p-nitrophenyl nitronyl nitroxide

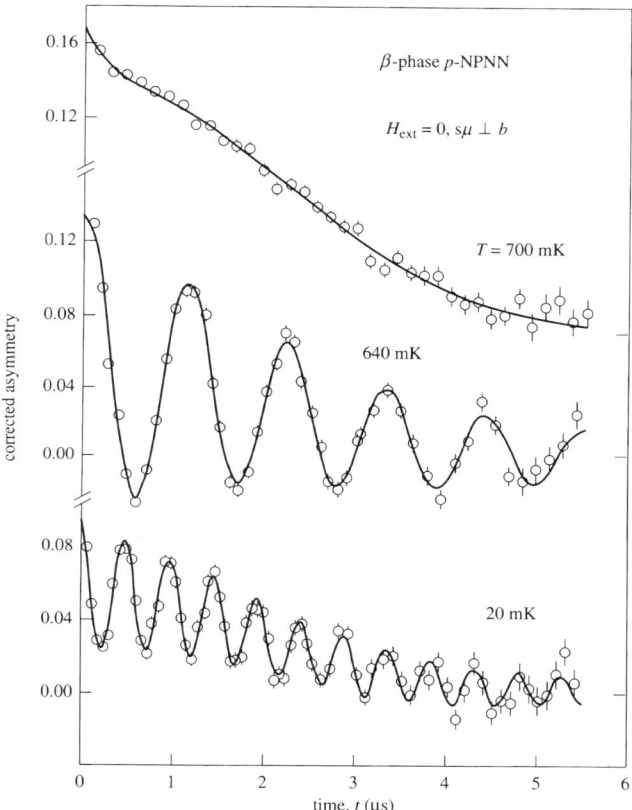

Fig. 17. The ZF-μSR spectra observed in the β-phase crystals of p-NPNN with initial muon spin polarized perpendicularly to the b-axis

ferromagnetism. One is the interaction within the ac-plane and the other is the interaction between the neighboring ac-planes. The former interactions, shown by the dotted lines in Fig. 19, form an approximate 2-D square lattice [15], while the latter interactions, shown by the solid lines, result in a distorted tetrahedral coordination [20, 26]. According to the calculation by Okumura et al., the former interaction is estimated to be 2 J_{12}/k_B = 0.48 K, and the latter to be 2 J_{13}/k_B = 0.22 K [27]. It has been suggested that the magnetic order in p-NPNN is due to a magnetic dipole interaction instead of the electronic exchange interaction [28]. However, our calculation of dipole-dipole interaction, D, based on the spin density data from the neutron diffraction experiments results in the values of D_a/k_B = −0.016 K, D_b/k_B = −0.029 K, and D_c/k_B = 0.045 K. These values indicate that the dipole interaction in the β-phase crystal stabilizes the spin system when the ferromagnetic spin alignment is assumed along the b-axis, but that it is smaller by one order of magnitude to

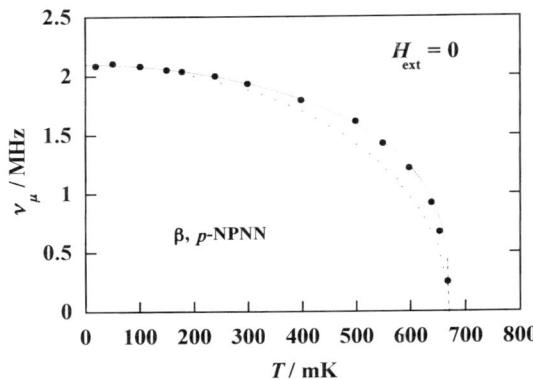

Fig. 18. The temperature dependence of muon spin precession frequency in the β-phase crystals. The frequency is proportional to the spontaneous magnetization. The solid curve is a fit to $M(T) = [1-(T/T_C)^\alpha]^\beta$ with $\alpha = 1.86$ and $\beta = 0.32$ and the *dotted curve* represents calculation for isotropic 3-D Heisenberg model

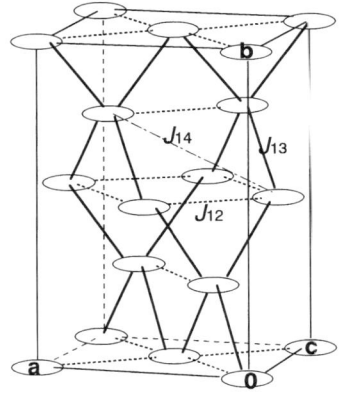

Fig. 19. Schematic drawing of the crystal structure of the β-phase crystal of p-NPNN. Each ellipsoid represents the p-NPNN radical molecule. The three kinds of exchange interactions, J_{12}, J_{13}, and J_{14}, are identified

account for the experimental T_C value. Therefore, we are of the opinion that the Curie temperature, T_C, is essentially governed by the electronic exchange interaction, but that the direction of the magnetic easy axis is determined by the dipole interaction.

5.6
Pressure Induced Ferro- to Antiferromagnetic Transition

When the pressure is applied to the crystal, the a.c. susceptibility exhibits remarkable changes. Figure 20 shows the pressure dependence of the a.c.

An Organic Radical Crystal Showing Spontaneous Ferromagnetic Order

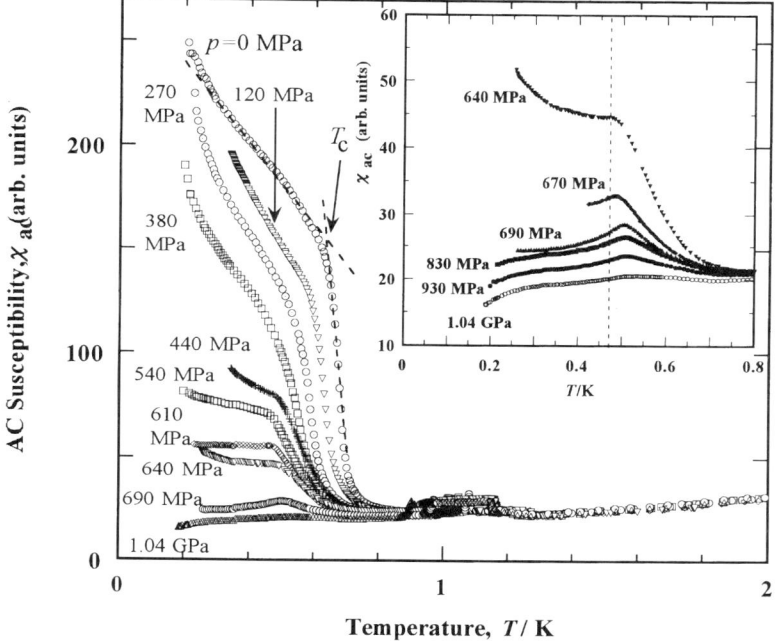

Fig. 20. The temperature dependence of a.c. susceptibility of the polycrystalline sample of the β-phase p-NPNN as a function of applied pressure. The crystals undergo a ferro- to antiferromagnetic transition at around 650 MPa as shown in the *inset*

susceptibility of the polycrystalline sample of β-phase p-NPNN up to $p = 1.04$ GPa measured under the a.c. field $H_{ac}(v) < 1$ Oe $(=10^3/4\pi$ A m$^{-1})$ with the frequency $v = 15.9$ Hz [29, 30]. It is found that the critical temperature, T_C, defined by the crossing point of the extrapolated lines from above and below T_C, approximately agrees with that determined from the heat capacity peak. As shown in Fig. 20, T_C shifts towards the lower temperature side with the initial gradient $d[T_C(p)/T_C(p_0)]/dp = -0.48$ GPa^{-1}, and the magnitude of the susceptibility decreases gradually as the pressure increases from p_0 (=0 MPa). In the low pressure region below $p \approx 650$ MPa, however, the ferromagnetic behavior is still preserved below $T_C(p)$, as characterized in the shape of χ_{ac}.

In the high pressure region above 650 MPa, in contrast, the magnitude of χ_{ac} becomes quite small, the susceptibility in the ordered state decreases sensitively with the pressure increase, and the shoulder-like curve of χ_{ac} around $T_C(p)$ changes into a cusp as shown in the inset of Fig. 20. Furthermore, $T_C(p)$ increases as the pressure increases with the gradient of $d[T_C(p)/T_C(p_c)]/dp = +0.04$ GPa^{-1}, where $p_c = 650 \pm 50$ MPa is the critical pressure. These results suggest that the magnetic order below T_C is that of an antiferromagnet under high pressure.

The antiferromagnetic behavior is also recognized in the external field dependence of χ_{ac} at constant pressure, $p = 690$ MPa as shown in Fig. 21 [29, 30]. $T_C(p)$ shifts to lower temperature as the field increases, in contrast to the case of a ferromagnet. Thus, we can conclude that pressurization induces a ferromagnetic to antiferromagnetic transition in the β-phase crystal.

5.7
Charge-Transfer Mechanism

These experimental results could be explained in terms of charge transfer mechanism mentioned in Sect. 3.3 by competition between the ferromagnetic and antiferromagnetic interactions. The effective exchange interaction between A and B molecules contains the kinetic (J_{AB}^K) and the potential (J_{AB}^P) terms as

$$J_{AB} = J_{AB}^K + J_{AB}^P \qquad (1)$$

Both terms depend on the overlap of molecular orbitals (MOs) on A and B molecules. The essential point of the charge transfer mechanism is that the kinetic exchange interaction is described by a sum of terms contributing to antiferro- and ferromagnetic interactions:

$$J_{AB}^K = -\frac{t_{SS}^2}{U} + \frac{t_{SF}^2}{U^2} J^{in} + \text{(terms related to other paths)} \qquad (2)$$

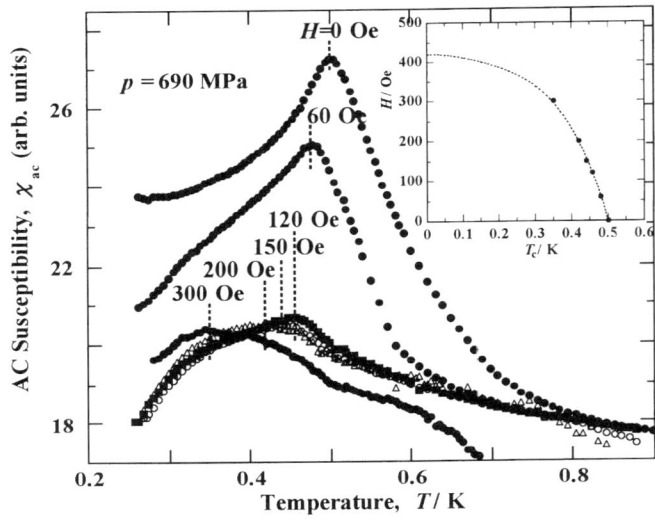

Fig. 21. The temperature dependence of a.c. susceptibility of the polycrystalline sample of the β-phase p-NPNN in the antiferromagnetic region under 690 MPa as a function of applied field. The fields corresponding to the susceptibility peaks are plotted against the critical temperature in the inset

where t_{SS} stands for the transfer integral between SOMOs of A and B molecules, t_{SF} for that between SOMO and other FOMOs, U is the on-site Coulomb repulsion, and J^{in} is the intramolecular exchange integrals. Then interplay or frustration among these contributions would result in $J_{AB} \approx 0$ in a certain condition, giving $T_C(p) \approx 0$ K. In the case of the β-phase of p-NPNN, we must take at least twelve interacting molecules adjacent to a central molecule in its zinc blende-like structure of Fig. 19. The corresponding exchange integrals are classified into three types J_{12}, J_{13}, and J_{14} from the symmetry of the lattice. As mentioned before, J_{12} and J_{13} are, from theoretical calculation, known to be ferromagnetic and J_{14} is only weakly antiferromagnetic. On the other hand, reduction of the dimensionality from the three- to two-dimensional ferromagnetic system has been pointed out from the appearance of short-range order effect by the heat capacity measurements under pressure [31]. From the crystal symmetry, it is obvious that only the exchange integral J_{12} is responsible for the two-dimensional ferromagnetic interaction. The transition temperature, T_{3D}, in such a reduced system, can be written in terms of the mean field theory as

$$k_B T_{3D} \propto S^2 \xi_{2D}^2 (J_{12}, T_{3D})\{|J_{13} + J_{14}|\}, \tag{3}$$

where ξ_{2D} is the spin correlation length in the ac-plane in which $J_{12}/k \approx 0.8$ K is estimated from the heat capacity curve [30, 31]. Therefore, the antiferromagnetic behavior at $p > p_c = 650 \pm 50$ MPa can be ascribed to a change in the sign of J_{13}. This means that the relative importance of the first and second terms in Eq. (2) relieve each other for J_{13} under high pressure.

5.8
Lattice Constants Under High Pressure

The change of the magnetic interactions discussed above would be closely related to changes in molecular shape and packing in the crystal. The lattice constants under various pressures have been determined by the Riedvelt method from the X-ray powder patterns obtained on the polycrystalline sample of β-phase p-NPNN by the use of imaging plates. The powder patterns observed are similar to one another, indicating that the crystal symmetry is maintained under pressure up to 1.26 GPa [30, 31]. The peak positions, of course, shift toward the direction of wider angles or to the direction of lattice contractions as the pressure increases. The lattice constants are plotted against pressure in Fig. 22. The crystal changes in two steps. The linear and volume compressibilities are summarized in Table 3. The biggest contraction (~4.5%) is found again along the c-axis as in the case of thermal contraction. The crystal density increases up to as high as 1.58 g cm^{-3} at 1.26 GPa.

The two-step contraction observed here could be understood in the following way. The molecular shape at 6 K is compared with that at room temperature in Fig. 23 [32]. The molecules are rotated by ± 3.3 ($\pi/180$) rad from the orientation at room temperature, the nitro groups being further rotated by ± 1 ($\pi/180$) rad. The lattice constants at 6 K (see Table 2) [25] nearly corresponds to those at about 400 MPa. From this, we could expect that the

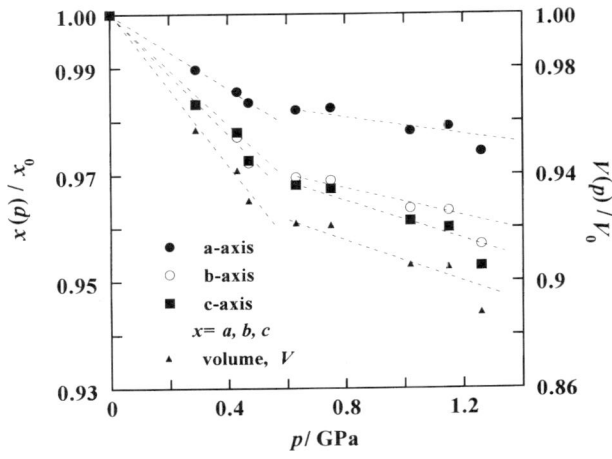

Fig. 22. The pressure dependence of the lattice constants of the β-phase crystal of p-NPNN

Table 3. Compressibility of the β-phase p-nitrophenyl nitronyl nitroxide crystal in units of Pa^{-1}

	$\kappa_a = -\frac{1}{a}\frac{\partial a}{\partial P}$	$\kappa_b = -\frac{1}{b}\frac{\partial b}{\partial P}$	$\kappa_c = -\frac{1}{c}\frac{\partial c}{\partial P}$	$\kappa = -\frac{1}{V}\frac{\partial V}{\partial P}$
$P < 550$ Pa	34.4×10^{-11}	56.6×10^{-11}	55.1×10^{-11}	143×10^{-11}
$P > 550$ Pa	11.5×10^{-11}	18.4×10^{-11}	22.3×10^{-11}	49.0×10^{-11}

molecules are librationally rotated about the a-axis to some further extent under pressures up to about 550 MPa. However, the plane of the nitro group has already been at the upright orientation at 6 K (or equivalently at about 400 MPa) with respect to the ac-plane, and further pressurization beyond 550 MPa would not increase the tilt angle any more for the nitro group. Then, the other parts of the molecule would rotate internally about the long molecular axis under higher pressures. Internal rotations of the five-membered ring having the bulky tetramethylethylene group would be the most probable candidates to take place in order for the molecule to become more and more planar and for the crystal to become more compact.

Such molecular deformation would cause a change in the electronic structure of the molecule and in the intermolecular magnetic interactions. At this moment we cannot conclude definitely, because the crystal structure at very low temperature and under high pressure is not available. However, it is natural to expect that the molecules take a more planar form and a more parallel arrangement in the crystal on compression at low temperature. The intermolecular charge transfer interaction between the SOMOs on adjacent molecules would become more efficient, resulting in interchange of the importance of ferromagnetic and antiferromagnetic interactions, as discussed in the preceding section.

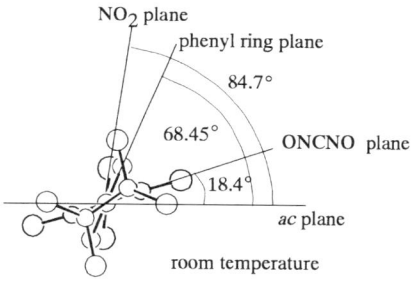

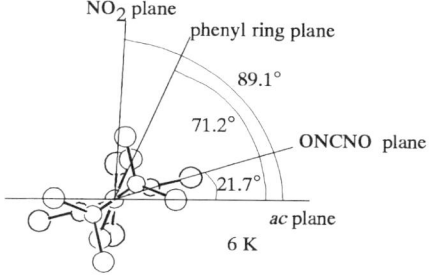

Fig. 23. The molecular shape of *p*-NPNN in the β-phase crystal at room temperature and at 6 K

6
Antiferromagnetism of γ-Phase Crystal of *p*-NPNN

It was initially suggested that the γ-phase crystal of *p*-NPNN can become a ferromagnet because the heat capacity exhibited a sharp peak and the a.c. susceptibility became quite large at 0.65 K [33]. However, it was later found that the heat capacity peak remains even in the magnetic field up to 1800 Oe, as is seen in Fig. 24 [34]. This is in contrast to the case of the β-phase and is characteristic of an antiferromagnet. Despite that, it was not established why the a.c. susceptibility became so large. Various efforts have made it clear that the phenomenon is induced by contamination of the ferromagnetic β-phase, to which the γ-phase crystal transforms during the measurements at low temperatures [34].

After various trials to determine the proper experimental conditions, it became possible to measure the a.c. susceptibility of the γ-phase itself with a powder sample. Figure 25 shows the results of a.c. susceptibility measurements with the powder sample. The curve at the zero field shows the characteristic feature of antiferromagnets, namely peaking at the transition temperature and decreasing to about 2/3 of the maximum in the low temperature side. Thus it is concluded that the γ-phase crystal of *p*-NPNN undergoes a transition to an AFM order at $T_N = 0.65$ K.

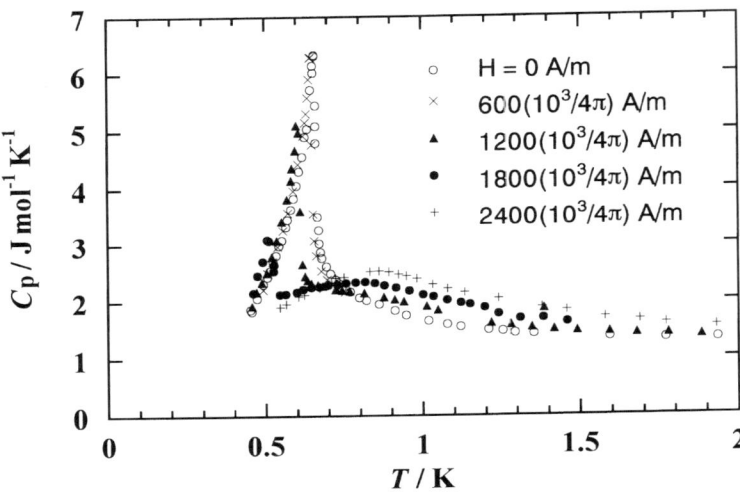

Fig. 24. The temperature dependence of heat capacity of the γ-phase crystal of p-NPNN as a function of applied field. Note the heat capacity peak persists up to the field as high as 1800 Oe

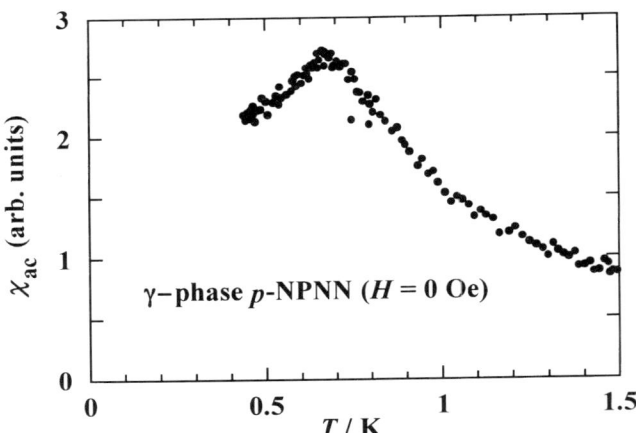

Fig. 25. The temperature dependence of polycrystalline sample of the γ-phase p-NPNN

It should be noted that the magnetic susceptibility and the heat capacity of the γ-phase are well explained by a 1-D isotropic FM Heisenberg model with $2 J/k_B = 4.3$ K in the temperature range above 4 K [35]. This FM interaction is believed to operate along the [0 1 1] direction. For the other two directions, FM coupling with $2 J'/k_B = 0.22$ K and AFM coupling with $2 J''/k_B = -0.18$ K are suggested, from the analysis of the field dependence of the heat capacity.

An Organic Radical Crystal Showing Spontaneous Ferromagnetic Order

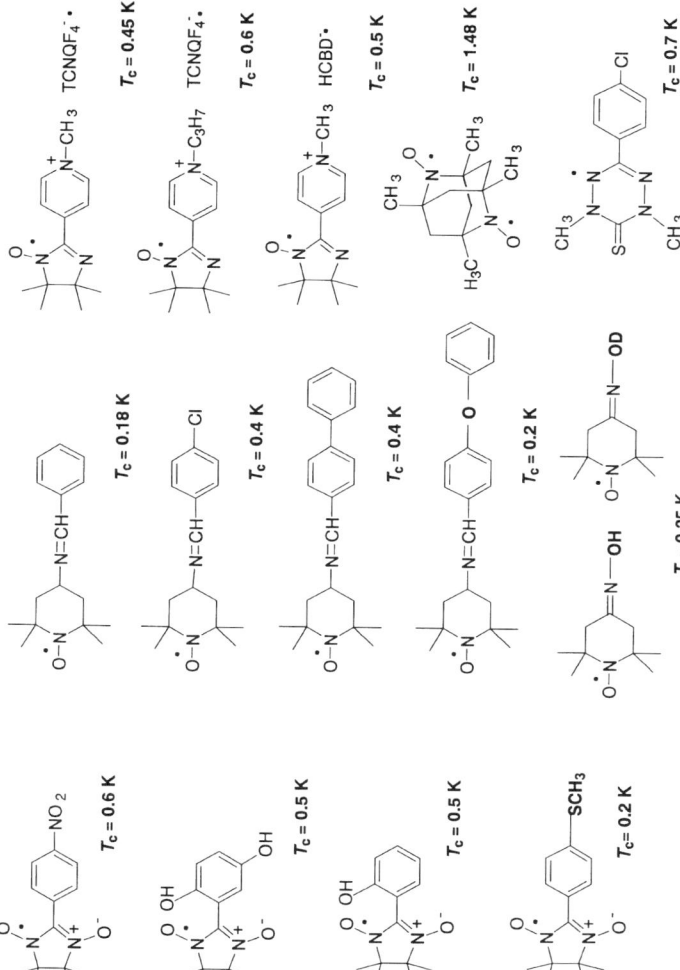

Fig. 26. Some organic ferromagnets reported

7
Other Ferromagnetic Organic Crystals

Following the finding of the ferromagnetism in p-NPNN, a dozen organic compounds have been found to become a ferromagnet [36-44]. These are shown in Fig. 26. Most substances undergo the ferromagnetic transition below 1 K. Diazaadamantane dioxide, which has the triplet ground state, exhibits the highest transition temperature of 1.48 K [38]. The transition temperature T_C is proportional to $S(S + 1)$ in the mean field approximation, and this means that the magnitude of the ferromagnetic interaction is nearly the same as that of the other compounds.

In order to increase the ferromagnetic transition temperature, it is indispensable to introduce other mechanisms other than those described here. Introduction of conductivity is one of the possibilities and various trials are being attempted in several laboratories.

8
References

1. Tamura M, Nakazawa Y, Shiomi D, Nozawa K, Hosokoshi Y, Ishikawa M, Takahashi M, Kinoshita M (1991) Chem Phys Lett 186: 401
2. Mukai K (1969) Bull Chem Soc Japan 42: 40
3. Awaga K, Sugano T, Kinoshita M (1986) Solid State Commun 57: 453
4. Awaga K, Sugano T, Kinoshita M (1986) J Chem Phys 85: 2211
5. Awaga K, Sugano T, Kinoshita M, Matson T, Suga H (1987) J Chem Phys 87: 3062
6. Kosaki A, Suga H, Seki S, Mukai K, Deguchi Y (1969) Bull Chem Soc Japan 42: 1525
7. Williams DE (1969) Mol Phys 16: 145
8. Awaga K, Sugano T, Kinoshita M (1986) Chem Phys Lett 128: 587
9. Awaga K, Sugano T, Kinoshita M (1987) Chem Phys Lett 141: 540
10. Awaga K, Sugano T, Kinoshita M (1988) Synth Met 27B: 631
11. Kinoshita M (1989) Mol Cryst Liq Cryst 176: 163
12. McConnell HM (1967) Proc RA Welch Found Chem Res 11: 144
13. McConnell HM (1963) J Chem Phys 39: 1910
14. Allemand P-M, Fite C, Canfield P, Srdanov G, Keder N, Wudl F (1991) Synth Met 41/43: 3291
15. Awaga K, Inabe T, Nagashima U, Maruyama Y (1989) J Chem Soc Chem Commun 1617, cf. (1990) 520. Care should be given to the axis system of the β-phase crystal in different references. Here the system in Table 2 is adopted, but other authors use different systems
16. Awaga K, Inabe T, Yokoyama T, Maruyama Y (1993) Mol Cryst Liq Cryst 232: 79
17. Turek P, Nozawa K, Shiomi D, Awaga K, Inabe T, Maruyama Y, Kinoshita M (1991) Chem Phys Lett 180: 327
18. Nakazawa Y, Tamura M, Shirakawa N, Shiomi D, Takahashi M, Kinoshita M, Ishikawa M (1992) Phys Rev B46: 8906
19. Awaga K, Maruyama Y (1989) Chem Phys Lett 158: 556
20. Kinoshita M (1993) Mol Cryst Liq Cryst 232: 1
21. Le LP, Keren A, Luke GM, Wu WD, Uemura YJ, Tamura M, Ishikawa M, Kinoshita M (1993) Chem Phys Lett 206: 405
22. Uemura YJ, Le LP, Luke GM (1993) Synth Met 56: 2845
23. Oshima K, Kawanoue H, Haibara Y, Yamazaki H, Awaga K, Tamura M, Ishikawa M, Kinoshita M (1995) Synth Met 71: 1821

24. Oshima K, Haibara Y, Yamazaki H, Awaga K, Tamura M, Kinoshita M (1995) Mol Cryst Liq Cryst 271: 29
25. Zheludev A, Ressouche M, Schweizer E (1994) Solid State Commun 90: 233
26. Awaga K, Inabe T, Yokoyama T, Maruyama Y (1993) Mol Cryst Liq Cryst 232: 79
27. Okumura M, Mori W, Yamaguchi K (1993) Mot Cryst Liq Cryst 232: 35
28. Miller JS, Epstein AJ (1994) Angew Chem Int Ed Engl 33: 385
29. Mito M, Kawae T, Takumi M, Nagata K, Tamura M, Kinoshita M, Takeda K (1997) Phys Rev B 56: 14,255
30. Takeda K, Mito M, Kawae T, Takumi M, Nagata K, Tamura M, Kinoshita M (1998) J Phys Chem 102: 671
31. Takeda K, Konishi K, Tamura M, Kinoshita M (1996) Phys Rev B 53: 3374
32. Kinoshita M (1991) Phil Trans R Soc Lond A 357: 2855
33. Kinoshita M, Turek P, Tamura M, Nozawa K, Shiomi D, Nakazawa Y, Ishikawa M, Takahashi M, Awaga K, Inabe T, Maruyama Y (1991) Chem Lett 1225
34. Takahashi M, Kinoshita M, Ishikawa M (1992) J Phys Soc Japan 61: 3745
35. Takahashi M, Turek P, Nakazawa Y, Tamura M, Nozawa K, Shiomi D, Ishikawa M, Kinoshita M (1991) Phys Rev Lett 67: 746
36. Nogami T, Tomioka K, Ishida T, Yoshikawa H, Yasui M, Iwasaki F, Iwamura H, Takeda N, Ishikawa M (1994) Chem Lett 29
37. Ishida T, Tsuboi H, Nogami T, Yoshikawa H, Yasui M, Iwasaki F, Iwamura H, Takeda N, Ishikawa M (1994) Chem Lett 919
38. Chiarelli R, Novek A, Rassat A, Tholence JL (1993) Nature 363: 147
39. Sugawara T, Matsushita MM, Izuoka A, Wada N, Takeda N, Ishikawa M (1994) J Chem Soc Chem Commun 1723
40. Mukai K, Nedachi K, Takiguchi M, Kobayashi T, Amaya K (1995) Chem Phys Lett 238: 61
41. Nogami T, Togashi K, Tsuboi H, Ishida H, Yoshikawa H, Yasui H, Iwasaki F, Takeda N, Ishikawa M (1995) Synth Met 71: 1813
42. Cirujeda J, Mas M, Molins M, Lanfranc de Panthou F, Laugier J, Park JG, Paulsen C, Rey P, Rovira C, Veciana J (1995) J Chem Soc Chem Commun 709
43. Nogami T, Ishida T, Tsuboi H, Yoshikawa H, Yamamoto H, Yasui M, Iwasaki F, Iwamura H, Takeda N, Ishikawa M (1995) Chem Len 635
44. Nakatsuji S, Morimoto H, Anzai H, Kawashima J, Maeda K, Mito M, Takeda K (1998) Chem Phys Lett 296: 159

The Mechanism of the Through-Space Magnetic Interactions in Purely Organic Molecular Magnets

Juan J. Novoa, Mercè Deumal

Departament de Química Física, Facultat de Química and CER Química Teòrica, Universitat de Barcelona, Av. Diagonal 647, 08028 Barcelona, Spain
E-mail: novoa@qf.ub.es

The mechanism of the intermolecular magnetic (or through-space) interactions found in purely organic molecular crystals is reviewed, using results from the *ab initio* studies of model systems, and from the statistical analysis of the packing of nitronyl nitroxide crystals presenting dominant ferro or antiferromagnetic interactions. First of all, the foundations of the McConnell-I mechanism (more properly called a proposal) are reviewed from a rigorous theoretical point of view. It is shown that this proposal lacks a rigorous foundation and works in some prototype systems due to error compensations associated to the high symmetry of the model systems employed to evaluate such a mechanism. It will be shown how the McConnell-I mechanism fails in rationalizing the magnetic properties of well known nitronyl nitroxide crystals. We will also show the existence of pitfalls in many of the magneto-structural correlations employed today. One example of an erroneous correlation is that which associates the magnetic character of a nitronyl nitroxide dimer with the relative orientation of the ONCNO groups in these two molecules. Consequently, new magneto-structural relationships are needed based on unbiased assumptions. For such a purpose, we need to have solid data on the properties of the through-space magnetic interactions at the microscopic level. We will review in the last sections the current knowledge about these properties, obtained from *ab initio* computations and crystal packing studies.

Keywords: Molecular magnetism, Nitronyl nitroxides, Though-space magnetic interactions, Magnetic interaction mechanism, McConnell-I mechanism, Quantum chemical *ab initio* studies, Statistical analysis

1	Introduction	34
2	The McConnell-I Mechanism of Through-Space Magnetism: A Theoretical Analysis	35
3	Failure of the McConnell-I Mechanism in Nitronyl Nitroxide Crystals	38
3.1	Evidence from Individual Crystal Studies	38
3.2	Evidence from *ab initio* Computations on Small Model Radical Dimers	42
3.3	Evidence from Statistical Magneto-Structural Studies in Nitronyl Nitroxide Crystals	43
4	*Ab initio* Study of the Nature of the Through-Space Magnetic Interactions Present in Nitronyl Nitroxide Crystals: The LICMIT and WILVIW10 Case	51

5	Concluding Remarks	57
6	References	59

1
Introduction

Molecular magnetism is a macroscopic property found in solids having long-living, spin-containing units (that is, stable molecular radicals, in the case of purely organic molecular magnets) oriented in such a way that they allow the propagation of the magnetic interactions throughout the solid [1]. In the case of purely organic molecular magnets, the magnetic interaction occurs mainly by direct overlap of the orbitals of the nearby units, in what are called *through-space interactions*, to distinguish them from the *through-bond interactions* found in organometallic solids, where the spin-containing metallic centers interact by means of the orbitals of the ligands placed between these centers (this is one example of the superexchange mechanism) [1a, b, 2].

The properties of the mechanism of the through-space magnetic interactions are not well understood at the present moment, although some basics facts are well established. Thus, for instance, the dependence of the magnetic interaction on the relative orientation of the radicals within the crystal is well known. Such a property is clearly demonstrated by the different magnetic properties found in different polymorphic forms of the same radical [3] (which, by definition, can only differ in the relative orientation of the same radicals). However, for a proper rationalization of the current molecular magnets or for a rational design of new, purely organic materials showing magnetic properties, one needs to go one step further and be able to identify which relative orientations of the radicals give rise to ferromagnetic interactions in the crystal, and which ones generate antiferromagnetic interactions. One should mention here that having good magneto-structural correlations is only half of the problem in designing molecular magnets. In addition, one has to learn how to control the packing of the radicals within the crystal [4, 5] to guarantee the presence of the desired magnetic interactions in the crystal. This is an example of what is nowadays called *crystal engineering* [6]. We will not discuss the complexities of crystal engineering in detail here. Instead, our aim is to review the current knowledge on the mechanism of the through-space magnetic interactions, at a microscopic level.

At the present time, there are some empirical magneto-structural correlations published in the literature, generated after looking for common factors in the relative orientations of the radicals whose crystals show dominant magnetic properties. Thus, for instance, it is commonly accepted that short N—O···O—N contacts in the nitronyl nitroxide crystals are indicative of dominant antiferromagnetic interactions between the dimers in which such contacts are found [7]. Besides these empirical correlations, the most popular magneto-structural correlation is the so called McConnell-I proposal (or mechanism) [8], which is based on theoretical considerations. It receives this name because it was the first of two proposals presented by McConnell to describe the mechanism of the through-space magnetism [9]. Originally introduced to rationalize the through-space magnetism in crystals whose packing motif are piles presenting π-π short contacts, it is nowadays used on any kind of molecular crystal. It is a very popular mechanism because it is easy to use: the rationalization of the crystal magnetic properties is done by looking at the products of the atomic spin populations of the atoms making the shortest intermolecular contacts between pairs of adjacent radicals [1].

For a long time, the McConnell-I mechanism was believed to work [1], in particular after its predictions had been successfully matched against the experimental results obtained on the [2.2]paracyclophane isomers [10]. However, there are also cases in which the McConnell-I predictions are in conflict with the experimentally measured magnetic properties. Here, we will illustrate some of these conflicting cases. Detailed theoretical and numerical investigations of the McConnell-I proposal showed that these failures are just a consequence of the deficiencies of the proposal: the basic expression in which the McConnell-I proposal is based is not theoretically sound [11]. Moreover, it was shown that the success of this proposal in the [2.2]paracyclophane isomers came from a fortuitous cancellation of errors, due to the high symmetry of these isomers [11]. Therefore, such a cancellation is not likely to take place in the general case.

Given this state of affairs, one needs to find new magneto-structural correlations based on sound principles. This can be done by investigating the microscopic properties of the through-space magnetic interactions along two lines of research:

(a) searching for general trends on the relative orientations of neighboring radicals in the known molecular magnets, by detailed statistical analysis of their crystal packings, and
(b) by quantum chemical computations on model systems, aimed at reproducing the type of magnetic interactions present in the purely organic molecular crystals.

Here we will present the main findings from investigations in which the two lines of research are used.

2
The McConnell-I Mechanism of Through-Space Magnetism: A Theoretical Analysis

In order to explain the intermolecular magnetism between *aromatic* radicals, in 1963 McConnell suggested [8] that the magnetic interaction between two aromatic radicals A and B could be approximated by the following Heisenberg spin Hamiltonian (Eq. 1)

$$\hat{H}^{AB} = -\sum_{i\in A, j\in B} J_{ij}^{AB} \hat{S}_i^A \hat{S}_j^B \tag{1}$$

in which J_{ij}^{AB} are two-center exchange integrals, whose form is (Eq. 2)

$$J_{ij} = [ij|ij] + 2\langle i|j\rangle\langle i|h|j\rangle \tag{2}$$

and $S_i^A S_j^B$ is the product of the *spin operators* of atoms i and j placed on fragments A and B, respectively. One should note that this expression is an approximation of the general Heisenberg Hamiltonian (Eq. 3) [12]

$$\hat{H}^{AB} = Q - \sum_{i,j} J_{ij} \left(2\hat{S}_i \hat{S}_j + \frac{1}{2}\hat{I}_{ij}\right) \tag{3}$$

in which the intrafragment terms have been neglected. Such an approximation is only valid if the expected values of the intrafragment components are the same for the states for which the general Hamiltonian acts.

McConnell [8] also proposed that the Hamiltonian of Eq. 1 could be replaced by a simplified form (Eq. 4)

$$\hat{H}^{AB} = -\hat{S}^A \hat{S}^B \sum_{i\in A, j\in B} J_{ij}^{AB} \rho_i^A \rho_j^B \tag{4}$$

In this expression the first two operators are the total spin operators for fragments A and B, J_{ij}^{AB} are same two-center exchange integrals defined above, and $\rho_i^A \rho_j^B$ is the product of the *atomic* spin density on atoms i and j of fragments A and B, respectively. This Hamiltonian is purely phenomenological, as there is no systematic set of approximations that derives it from Eq. 1 in a rigorous way [11].

One can gain a better insight into the properties of the McConnell-I Hamiltonian (Eq. 4) by computing energy differences between two states. If only one unpaired electron is present in each fragment, these electrons can be paired into a singlet (S) and a triplet (T) state. Using the following expression (Eq. 5)

$$\langle \hat{S}^A \hat{S}^B \rangle = \tfrac{1}{2}[S(S+1) - S_A(S_A+1) - S_B(S_B+1)] \tag{5}$$

it is possible to demonstrate that the energy difference between the singlet and triplet states obtained using the Hamiltonian of Eq. 4 is equal to Eq. 6 [11]

$$E^S - E^T = \langle \hat{H}^{AB}\rangle^S - \langle \hat{H}^{AB}\rangle^T = \sum_{i\in A, j\in B} J_{ij}^{AB} \rho_i^A \rho_j^B = J \tag{6}$$

where J is an effective coupling constant between the two states of interest.

According to this equation, it can be concluded that *the McConnell-I mechanism predicts that a triplet ground state is obtained in the interaction between two doublet fragments when the atoms making the shortest contacts present atomic spin populations of opposite sign* (as J_{ij}^{AB} is normally negative). Such an interpretation rests on the assumption that the values of the J_{ij}^{AB} integrals are determined mainly by the intermolecular atom-atom distance and are all similar. However, given the complexities of the molecular orbitals in which the interacting electrons are located, this is not necessarily the case and these integrals could be affected by other factors like the symmetry of the orbitals, or the relative position of the atoms in their fragments, for instance. Therefore, it is possible that the shortest contacts are not those presenting the dominant J_{ij}^{AB} values. As a consequence, the association is not always valid [11].

Another weak point of the "triplet-opposite signs" interpretation comes from the validity of Eqs. 4 and 6. As mentioned above, Eq. 4 is a simplification of the general case, whose validity has been never demonstrated. In fact, the derived expression for the $E^S - E^T$ energy difference obtained using the general Heisenberg Hamiltonian (Eq. 3) is different to that obtained with the simplified form (Eq. 7) [11]

$$E^S - E^T = \sum_{i,j} J_{ij}\left(P_{ij}^S - P_{ij}^T\right) \quad (7)$$

where P_{ij}^S and P_{ij}^T are the singlet and triplet exchange density matrices obtained from the singlet and triplet wave functions, using the following Eq. 8

$$P_{ij} = \left\langle -\left(2\hat{S}_i\hat{S}_j + \frac{1}{2}\hat{I}_{ij}\right)\right\rangle \quad (8)$$

Therefore, it is clear that the McConnell-I relationship is valid only if one can make the following association (Eq. 9):

$$\rho_i^A \rho_j^B \Leftrightarrow P_{ij}^S - P_{ij}^T \quad (9)$$

However, there is no reason why the product of atomic spin densities should be related to the difference of exchange density matrices. In the best case, this association has to be tested numerically before accepting it.

In order to test the applicability of Eq. 9, we carried out numerical computations on simplified models of the *ortho-*, *meta-*, and *para-*isomers of the bis(phenylmethylenyl)[2.2]paracyclophanes [10] (see Fig. 1), which represent their main structural and electronic properties. These computations were carried out by solving the exact Heisenberg Hamiltonian in a Valence Bond configurational space [11], using the so called MMVB method [13]. This method uses the MM2 potential to describe the inert σ-bonded framework, and a Heisenberg Hamiltonian to represent all the electrons involved in the π-conjugation or in the formation of new σ-bonds. The Heisenberg Hamiltonian has been parametrized to reproduce the results from accurate CASSCF computations on various model systems [13].

The ground states for the three model paracyclophane isomers agree well with the experimental results and the McConnell-I predictions. An analysis of the energetic components in these computations indicates that in the three isomers the dominant contribution to the singlet-triplet energy difference is that coming from the atoms making the shortest contacts, as one would expect if the McConnell-I proposal would work. The reason for this domination is the small numerical value of the J_{ij}^{AB} integrals associated to the

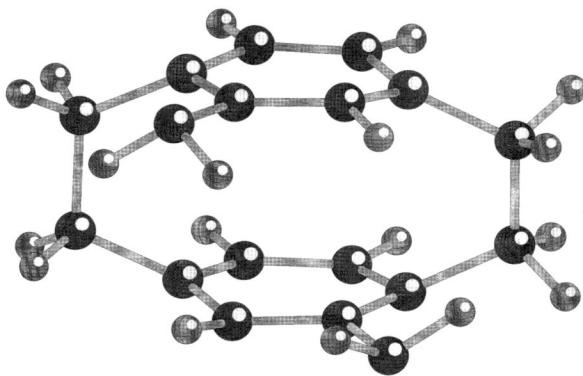

Fig. 1. Structure of the model employed to evaluate the properties of the [2.2]paracyclophanes in our MMVB computations. The structure corresponds to the *ortho* isomer

other terms present in Eq. 7, due to the high symmetry of the system. Therefore, on systems where such a high local symmetry environment is destroyed and the atoms are not well aligned, one should not expect that this cancellation will hold. Test computations on distorted paracyclophanes support such a conclusion. Consequently, *there is no reason to expect that the McConnell-I proposal will hold in general*. Our computations also showed that the J_{ij}^{AB} integrals have strong orientational and directional properties. Thus, even if the proposal would hold, a precise determination of the singlet-triplet stability cannot be done without a computation of the J_{ij}^{AB} integrals, in contrast to common practice which makes the predictions solely on the basis of the atomic spin populations.

3
Failure of the McConnell-I Mechanism in Nitronyl Nitroxide Crystals

3.1
Evidence from Individual Crystal Studies

Even if not all the interactions present in the crystals of purely molecular radicals are of the π-π type, the McConnell-I mechanism has been the main tool used in the analysis of the magnetic interactions present in molecular crystals. However, there is now enough evidence indicating that the McConnell-I predictions are not always valid. This is the case for many magnetic crystals of the nitronyl nitroxide family, whose general formula is indicated in Fig. 2 [1]. Here we will describe some of the evidence indicating the failure of the McConnell-I mechanism.

A first evidence against the validity can be obtained when analyzing the magnetic properties of many nitronyl nitroxides with well defined structures. We can illustrate the failure with one of these examples, the phenylnitronyl nitroxide crystal (PhNN, see Fig. 3) [14]. The X-rays structure of this crystal

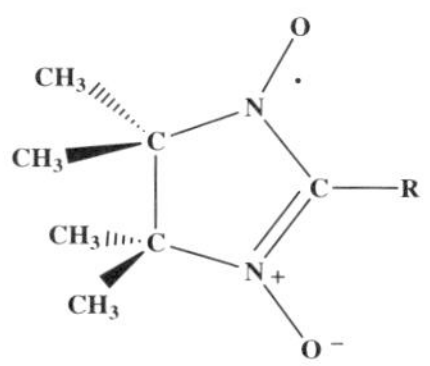

The Mechanism of the Through-Space Magnetic Interactions

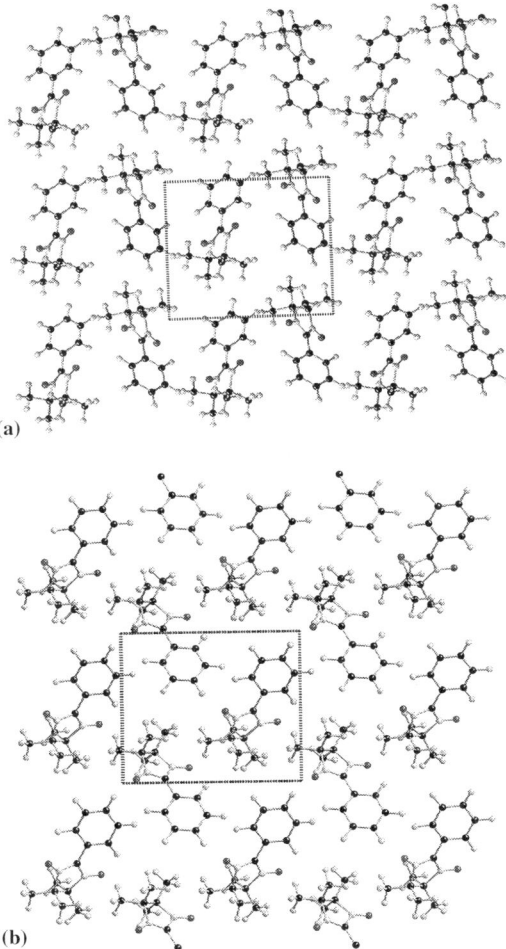

Fig. 2. Chemical structure of a general nitronyl nitroxide radical (R is the substituent which changes from one radical to another in this family)

Fig. 3. a Structure of the A plane of the PhNN crystal. **b** Structure of the B planes of the PhNN radical. Both planes extend along the crystallographic *b-c* plane, and are visualized along the *a* axis

has been determined with precision [14], and so was the spin density distribution using polarized neutron diffraction [15]. The crystal belongs to the $P2_1/c$ symmetry group, with cell parameters $a = 21.14$ Å, $b = 10.14$ Å, $c = 12.122$ Å, and $\beta = 108.1°$, and eight molecules per unit cell. The radical packs building planes on the b-c directions, which pile up in the a direction. There are two types of planes, the A and B planes shown in Fig. 3. Four planes pile up within each unit cell in an ABBA sequence. The experimental spin distribution agrees well (see Fig. 4) with that computed using the BLYP non-local exchange and correlation density functional [16, 17] and the 6-31G(2d,2p) basis set [18]. These results show that 90% of the spin density is located on the ONCNO atoms of the five-membered ring of the PhNN molecule (Fig. 4), being positive on the N and O atoms (0.29 and 0.35 e$^-$, respectively), and negative on the central C atom (-0.22 e$^-$). The remaining 10% of the spin density is distributed on the other atoms, with atomic spin populations in each atom smaller than 0.03 e$^-$ in absolute values. There is sign alternation between adjacent C atoms of the aromatic ring, and also between the N and C atoms of the five-membered ring. The amount of atomic spin population on the hydrogen atoms is always smaller than 0.003 e$^-$ in absolute values. It is important to mention here that the atomic spin population is not always proportional to the spin density at the nucleus, as obtained from the NMR or EPR hyperfine coupling constants (hfccs) [19]. This fact is particularly important on some of the hydrogens of the PhNN radical, and is inherent to the fact that the atomic spin population is an integration of the spin density on the nucleus and the rest of the space associated to that atom [20].

The PhNN crystal presents dominant antiferromagnetic properties [14]. However, according to the McConnell-I mechanism one should expect this crystal to be a paramagnet. This conclusion is reached after analyzing all the direct X$\cdots$Y short contacts between the atoms of the ONCNO group. Note that when extending the McConnell-I mechanism to σ interactions the short H$\cdots$O or H$\cdots$C contacts are disregarded because the atomic spin populations

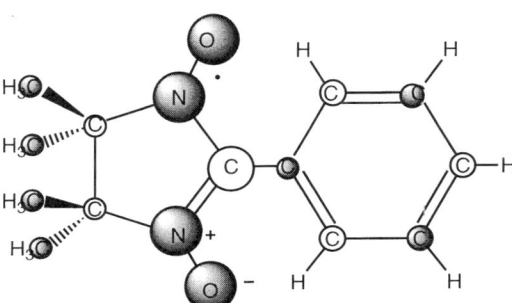

Fig. 4. Representation of the atomic spin population found in the atoms of the PhNN radical. Shaded atoms represent positive values, and unshaded negative ones. The size of the circles reflect the amount of atomic spin population (larger circles mean more population). The population on the H atoms is too small to be drawn

located in the H atoms are two orders of magnitude lower than the values in the ONCNO atoms. The shortest contacts between ONCNO groups within the A and B planes are nearly collinear O···O contacts located at 5.02 Å and 6.65 Å. Thus, given the large distance of these contacts, paramagnetism within the A and B planes should be expected if McConnell-I works properly. The situation for the interplane contacts is similar: the shortest ONCNO···ONCNO contacts found in between adjacent A-A planes is 4.51 Å, which becomes 4.05 and 3.69 Å when looking between adjacent A-B and B-B planes, respectively. The A-A and A-B distances are too large to allow an effective interplane magnetic interaction to be propagated all over the crystal. Therefore, as already mentioned, the McConnell-I predicts that the PhNN crystal should behave as a paramagnet, if one looks at the direct overlap of the ONCNO atomic spins.

The failure of the McConnell-I mechanism to predict magnetic properties of the PhNN crystal in good agreement with the experimental data, is only one of the many cases found in the nitronyl nitroxide crystals. Among them, there are various hydrogen bonded crystals whose short contacts are always of the A—H···B hydrogen bonded type. In such crystals, usually one finds that the ONCNO···ONCNO distances are very large and the McConnell-I mechanism systematically predicts a paramagnetic behavior. Therefore, to be able to explain the existence of the ferromagnetic or antiferromagnetic behaviors detected experimentally, the existence of magnetic interactions throughout the hydrogen bond is postulated [21]. This is a clear transgression of the McConnell-I mechanism. However, the existence of through-hydrogen bond magnetic interactions on hydrogen-bonded radical dimers is supported by theoretical computations on H_2NO dimers [22].

When one analyzes the magnetism of the PhNN crystal after allowing for the existence of magnetism through the C—H···ON contacts, the crystal is no longer a paramagnet: each PhNN molecule makes many C—H···ON contacts with other adjacent radicals in all directions of the space, due to the fact that the C—H···ON contacts are the strongest attractive contacts in the PhNN crystal in conjunction with the C—H···π contacts [5b]. Therefore, if one accepts that the McConnell-I mechanism can be extended to the C—H···ON contacts, and leaves aside the small value of the atomic spin population found in all hydrogens (this is equivalent to saying that the J_{ij} integrals for these interactions are stronger), it is possible to know if the contact is ferromagnetic or antiferromagnetic looking at the sign of the atomic spin population located on the hydrogen of the C—H group. A simple qualitative approach to define the sign of the atomic spin population on such hydrogens is to assign to that atom the opposite sign to that found in the carbon of the C—H group (see Fig. 4 for the atomic spin populations). Using this approach, all $C(sp^3)$—H···ON contacts should be ferromagnetic, while the magnetic character of the $C(sp^2)$—H···ON contacts depends on the aromatic C atom: H atoms attached to the C_{meta} atoms give rise to ferromagnetic $C(sp^2)$—H···ON contacts, while these contacts are antiferromagnetic in the C_{ortho} and C_{para} case. The overall nature of the interaction is difficult to define, as there are many contacts of opposite character within hydrogen-bonded crystals. Further studies are under way to determine the importance of these interactions in real crystals.

3.2
Evidence from *ab initio* Computations on Small Model Radical Dimers

The results of the previous section show the existence of inconsistencies between the experimental magnetic character and the predictions of the McConnell-I mechanism. However, these studies do not give information on the cause of the failure. One form of obtaining such information is by comparing the McConnell-I predictions on simple radical dimers with the results obtained from accurate *ab initio* computations on the same dimers.

We started our investigation on model dimers, by comparing the *ab initio* and McConnell-I predictions for the ground state of an H_2NO dimer oriented in such a way that the shortest contacts between the fragments are established between the oxygen atom of each fragment (Fig. 5). We made a potential energy surface scan in which the relative position of one fragment respect to the other was modified while keeping the O···O distance fixed at 3.0 Å, also forcing the N—O axis of one of the H_2NO fragment to be collinear with the oxygen atom of the other fragment. This scan was done while keeping all atoms in the plane of the fixed fragment (the *a-b* plane of Fig. 5). Alternatively, the NO group of the second molecule can be moved along the *a-c* plane of Fig. 5, while the hydrogens are in symmetrical position above and below that plane. In the two sets of geometries, the McConnell-I mechanism predicts that the interaction between these radicals must be antiferromagnetic, as the density in the two O atoms has the same sign (in this case, positive, like the N atom).

The singlet-triplet energy difference of the H_2NO dimers computed at the CASSCF(6,4)/6-31G(d) level for angles in the 90–180° range are shown in Fig. 5 [23]. By CASSCF(6,4)/6-31G(d) level one indicates computations done using the CASSCF method with a large (6,4) complete active space (that is, built by putting four electrons into six orbitals in all possible ways) and the 6-31G(d) basis set. We tested that the singlet-triplet energy separations did not change when the size of the active space or the basis set was increased. Therefore, *there are regions of the dimer in which the singlet is the ground state and others in which it has a triplet ground state*. This dual behavior depending on the relative orientation, goes against the predictions obtained using the McConnell-I model in the usual form, that is, by looking only at the atomic spin populations of the overlapping atoms.

The disagreement between the McConnell-I predictions and the *ab initio* results found in the H_2NO dimer can be extended to other systems, and, in particular, to the methyl···allyl radical dimer. Here, while the McConnell-I predictions are correct when the plane of the allyl radical is parallel to the plane of the methyl radical [23], it fails to predict the rapid change from ferromagnetic to antiferromagnetic when the methyl plane is tilted off the parallel arrangement.

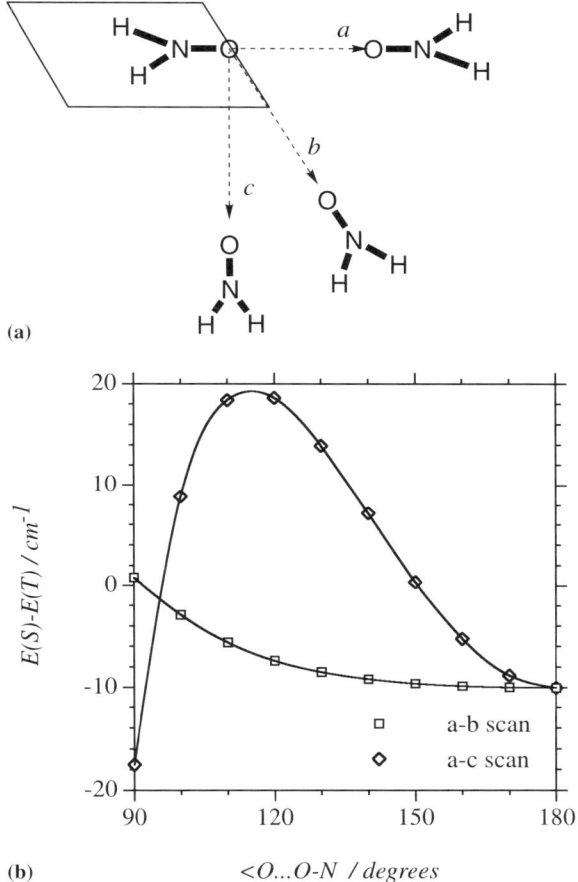

Fig. 5. a Diagram representing the orientation of the H$_2$NO···ONH$_2$ dimers in the scans along the *a-b* or *a-c* planes (see text for details). **b** Energetic difference between the singlet and triplet states as a function of the O···O—N angle

3.3
Evidence from Statistical Magneto-Structural Studies in Nitronyl Nitroxide Crystals

The previous two examples (Sects. 3.1 and 3.2) could be taken as exceptional cases, which, therefore, do not reflect the overall trend found in the experimental crystals. Therefore, it seems appropriate to investigate if it is possible to correlate, in a general statistical form, the magnetic properties of the crystal with the relative geometrical orientation of the adjacent ONCNO groups present in the nitronyl nitroxide crystals. According to the McConnell-I mechanism, crystals presenting dominant ferromagnetic interactions should present a different geometrical arrangement of the ONCNO groups than those found in crystals presenting dominant antiferromagnetic interactions. We thus selected those crystals presenting *dominant* magnetic properties, either ferro- or antiferromagnetic, because in this way we can assume that all their ONCNO···ONCNO short contacts exhibit the same magnetic character. The study was done on the nitronyl nitroxide crystals because they include many examples of crystals having dominant magnetic interactions, besides the fact that these crystal are the most interesting ones to analyze, because they include most of the purely organic magnets found up to now.

For our statistical study of the geometry of the ONCNO···ONCNO short contacts [24], we created a subset of nitronyl nitroxide crystals presenting dominant ferromagnetic interactions (the FM subset), and another separate subset with those crystals presenting dominant antiferromagnetic interactions (the AFM subset) [25]. These subsets were made by combining the data on nitronyl nitroxide crystals deposited in the Cambridge Crystallographic Database, with others provided by various authors. Of the initial set of crystals, we discarded those having R factors larger than 0.10 or having large molecular distortions. We also discarded co-crystals or crystals containing transition metals, and those crystals not showing dominant magnetic interactions in the 2–300 K range. We ended up with a database of 47 crystals, 23 of them in the dominant FM subset, and the remaining 24 in the AFM subset. The structures of the molecules included in each subset, with their refcodes (when available, otherwise we have assigned to them an internal refcode which begins by a 0) are shown in Figs. 6 and 7.

We then analyzed the relative orientation of the ONCNO···ONCNO short contacts using as parameters the six coordinates depicted in Fig. 8. Other studies [26] have shown that the internal geometry of these groups is nearly invariant among the crystals of this family. Therefore, under these circumstances, the six geometrical parameters of Fig. 8 are sufficient to fully describe the relative orientation of two ONCNO groups.

The results from this statistical analysis [24] show that many commonly accepted magneto-structural correlations are not valid. The first one is that short NO···ON contacts are indicative of dominant antiferromagnetic interactions in the crystal, a behavior consistent with the McConnell-I predictions. As shown in Table 1, short NO···ON contacts are found in the ferro- and antiferromagnetic subsets and in a similar number. In particular, in the 3–4 Å range of O···O distances, the number of contacts in each subset, 42% and 58%, is close to the 50% proportion that one should expect if the packing of the subsets of crystals would be identical.

Another correlation that does not hold is that associating the magnetic properties with the relative disposition of the ONCNO groups. If this association would work, there should be specific orientations of the ONCNO groups only present in the ferromagnetic subset, and similarly for the antiferromagnetic subset. In order to see such a failure, one has to include in the analysis the angular parameters (that is, A_1, A_2, T_1, T_2, and T_3). Figure 9 plots the position of the O_{21} atom relative to the $O_{12}-N_{12}-C_1-N_{11}-O_{11}$ atoms (which depends only on the D, A_1, and T_2 parameters) for the ONCNO···ONCNO contacts having an O···O distance shorter than 5 Å. It is clear from these plots

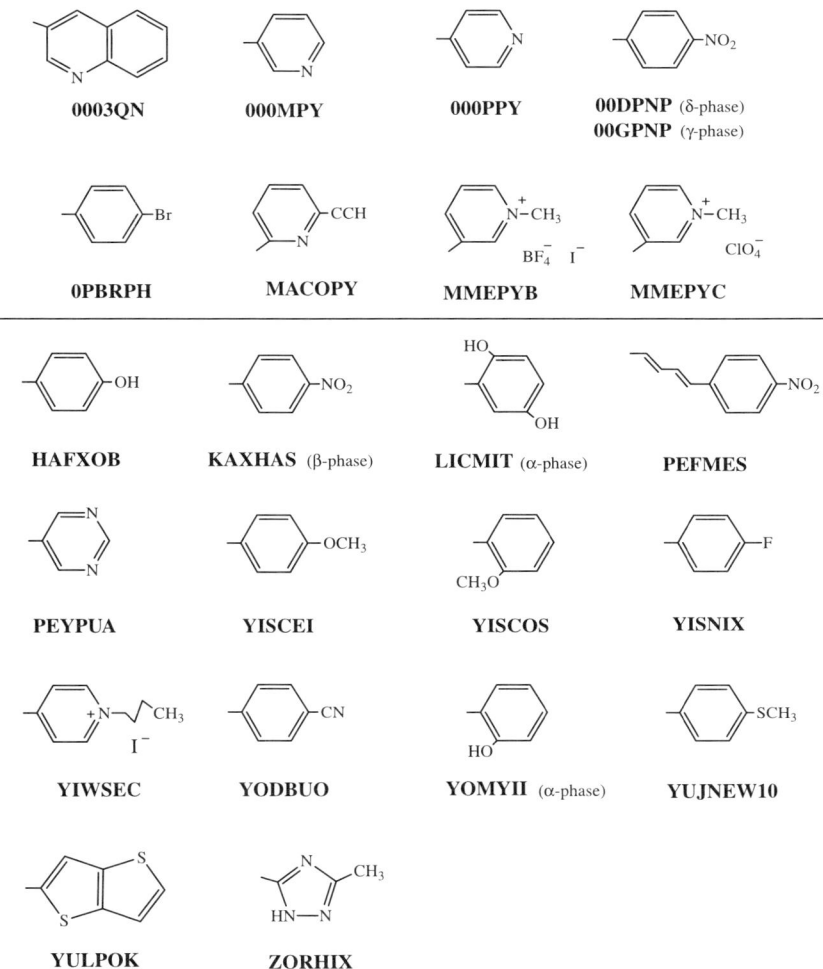

Fig. 6. Chemical structures of the R substituent for all the nitronyl nitroxide radicals whose crystals are included in the ferromagnetic subset

that the distribution of the O_{21} atom does not present regions in the FM subset excluded in the AFM subset, and vice versa. Instead, the distribution spreads over the whole range of values plotted, in a similar form. In fact, a cluster analysis of the sets of positions of the FM plus AFM subsets showed that the relative orientations of the FM subset are interlocked with those from the AFM subset, thus making impossible to distinguish them by the geometry of their contacts, that is, *there are no specific orientations of the ONCNO groups associated to a type of magnetic interaction*. The same conclusion was obtained when the range of O···O contacts included in the analysis was increased up to 10 Å.

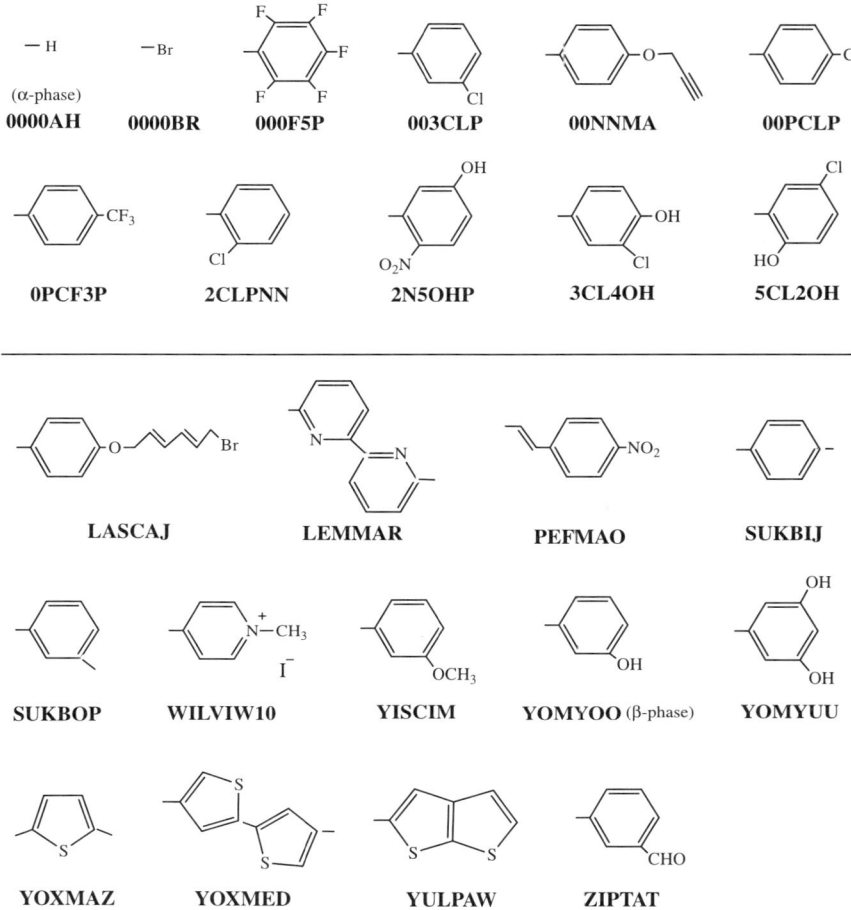

Fig. 7. Chemical structures of the R substituent for all the nitronyl nitroxide radicals whose crystals are included in the antiferromagnetic subset

When a similar analysis packing analysis is carried out for the shortest C—H···O—N contacts, the same conclusions are reached [24]: *there are no statistically significant differences in the relative dispositions of the shortest C—H···O—N contacts in the FM and AFM subsets of crystals.* Therefore, by combining the results for the O—N—CN—O···O—N—C—N—O and the C—H···O—N contacts, one can safely conclude that *it is not possible to determine the nature of the dominant magnetic interaction in a crystal by looking at the geometry of only one of these two types of contacts.*

The impossibility to generate a structure-correlation relationship based on the geometrical properties of one class of contacts is indicative of one of the following options:

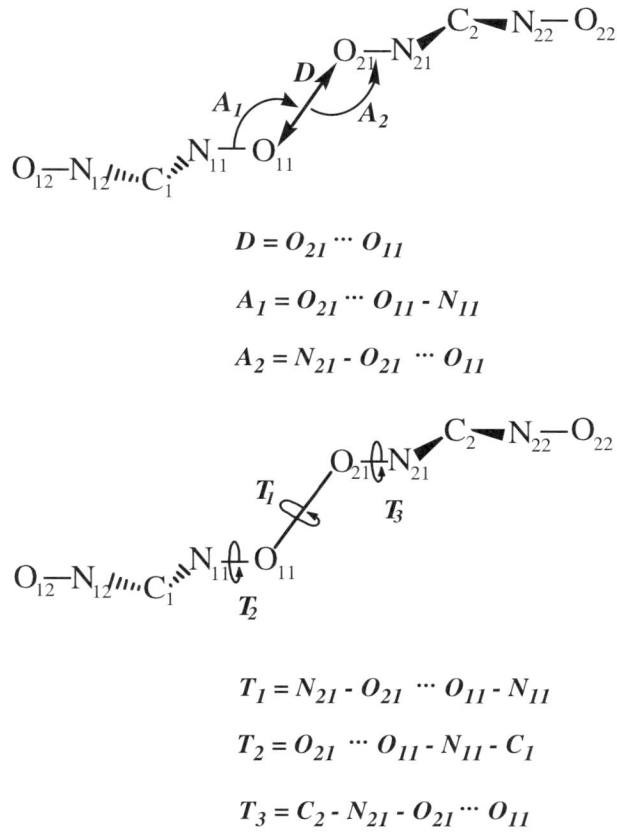

Fig. 8. Definition of the coordinates used to define the relative position of two ONCNO fragments when the internal geometry of these fragments is frozen. D is a distance, A_1 and A_2 are angles, while T_1, T_2 and T_3 are dihedral angles

(a) the McConnell-I is not working or is an oversimplification of the real mechanism, or
(b) the magnetic nature of the interaction between adjacent dimers is determined by the combined effect of more than one class of contacts and, consequently, it is not possible to obtain a reasonable magneto-structural correlation by looking only at one of the components.

The results presented in the previous sections indicate that the first option is certainly a valid one, but we will also explore if the second option is possible.

There is a final point to be stressed about this statistical analysis: the similarity between the geometry of the ONCNO···ONCNO contacts found in the FM and AFM subsets, also found for the C—H···ONCNO contacts. This is clearly manifested in the results of Tables 2 and 3, in which we collect the

Table 1. List of ONCNO···ONCNO contacts for crystals of the FM and AFM subsets within the range of distances indicated. Percentages of cases with intermolecular ferro- and antiferromagnetic interactions are also given

Distance range (Å)	FM subset			AFM subset	
	Total number of contacts	Number of contacts	%	Number of contacts	%
[0,3]	0	0	0	0	0
[0,4]	24	10	42	14	58
[0,5]	92	36	39	56	61
[0,6]	204	90	44	114	56
[0,7]	378	167	44	211	56
[0,8]	608	274	45	334	55
[0,9]	901	416	46	485	54
[0,10]	1312	611	47	701	53

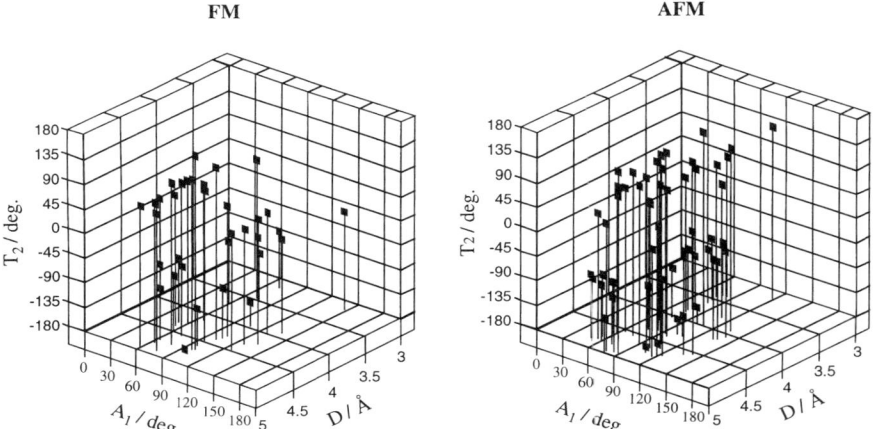

Fig. 9. Scatterplot of the values of the D, A_1 and T_2 parameters for all ONCNO···ONCNO contacts found in the FM and AFM subsets whose contacts lie between 3 and 5 Å. The scattergram in the left plots the FM subset contacts, and that in the right the AFM subset contacts

shortest contacts of each type found in the FM and AFM subsets. Clearly, there is a striking similarity in the distribution of distances and angles. This similarity is also found when one compares two crystals having the same range of distances. Thus, for instance, when we compare the two crystals of the FM and AFM subsets in which the shortest NO···ON contacts are found, the LICMIT [27] and WILVIW10 [28] crystals, the geometry of the shortest ONCNO···ONCNO contacts is very similar. This is illustrated in Fig. 10, where we have plotted the position of the dimers in which the shortest

Table 2. Values of the geometrical parameters defining the ONCNO···ONCNO contact for the five crystals of the FM and AFM subsets showing the shortest D distances. The distances are given in Å and the angles in degrees. The *refcodes* of crystals in which these interactions are found are given in Figs. 6 and 7

Refcode	D	A_1	A_2	T_1	T_2	T_3
FM subset						
LICMIT	3.158	146.8	146.8	180.0	−0.1	0.1
ZORHIX	3.168	126.9	71.5	119.0	46.5	−76.2
MMEPYC	3.429	153.2	68.0	173.3	−40.1	70.1
000MPY	3.499	127.5	76.4	−162.1	104.2	−92.1
PEYPUA	3.719	114.6	59.0	172.9	−114.0	−66.5
AFM subset						
WILVIW10	3.159	117.2	134.1	139.0	134.3	77.9
5CL2OH	3.369	82.6	82.6	180.0	75.6	−75.6
0000AH	3.431	120.0	75.8	−38.9	75.1	62.7
SUKBOP	3.522	77.7	77.7	180.0	−66.7	66.7
ZIPTAT	3.589	79.2	79.2	180.0	−82.7	82.7

Table 3. Values of the geometrical parameters defining the $C(sp^3)$—H···O—N contacts for the five crystals of the FM and AFM subsets showing the shortest D distances. The distances are given in Å and the angles in degrees. The *refcodes* of crystals in which these contacts are found are also indicated

Refcode	D	A_1	A_2	T_1
FM subset				
HAFXOB	2.339	137.9	167.2	−145.3
YOMYII	2.382	111.9	162.2	94.9
000MPY	2.419	118.0	168.6	160.5
ZORHIX	2.582	126.8	148.1	169.2
00GPNP	2.600	124.1	132.1	−150.9
AFM subset				
0000AH	2.295	88.9	158.0	54.5
YISCIM	2.343	162.6	161.5	−109.3
SUKBIJ	2.442	142.1	161.2	−148.0
0000BR	2.447	126.3	149.6	112.3
YOXMED	2.489	131.3	157.7	−157.8

ONCNO···ONCNO contacts are found. Notice that there are two dimers in Fig. 10 for the WILVIW10 crystal, the reason being that there is a second short contact at an O···O distance of 3.384 Å, besides the shortest 3.158 Å contact. All the other short ONCNO···ONCNO contacts in the LICMIT and WILVIW10 crystals are located at distances larger than 4.3 Å. The analysis of the dimers of Fig. 10 shows that the nearby ONCNO groups are oriented in a similar form in the two crystals, with these groups in a parallel arrangement: one of them displaced relative to the other in such a way that the shortest interfragment atom-atom distance between these ONCNO groups is a lateral O···O contact

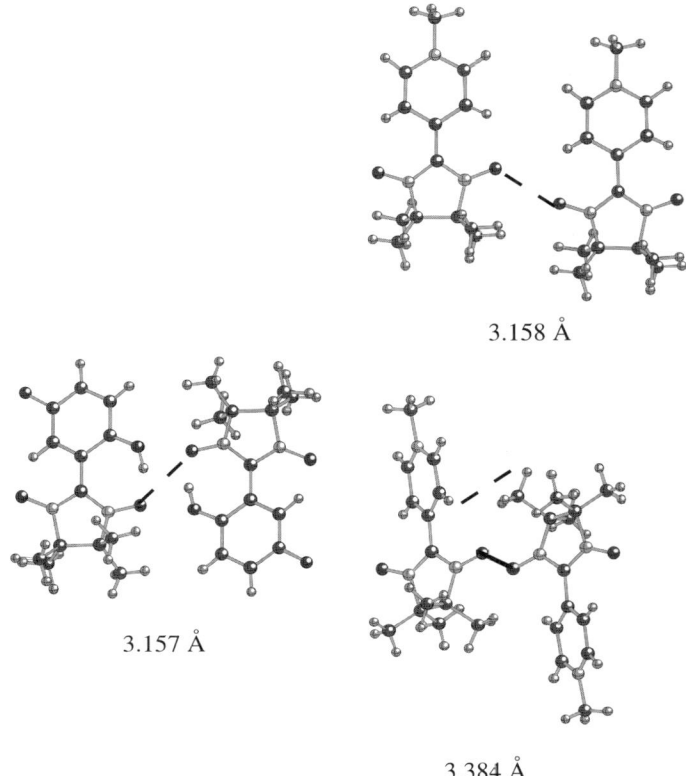

Fig. 10. Geometry of the dimers in which the shortest ONCNO···ONCNO contacts are found in the LICMIT (leftmost dimer) and WILVIW10 (two dimers in the right) crystals. The value of the O···O distance is also indicated

(see Fig. 11 for tridimensional views of the positions of the ONCNO groups). The fact that in the LICMIT crystal the radicals of the dimer are in an up-down disposition while in the WILVIW10 crystal they are in an up-up disposition, is totally irrelevant from the point of view of the McConnell-I mechanism. Furthermore, not all the crystals of the AFM subset presenting short NO···ON contacts have dimers packed in an up-up disposition. For example, the 5CL2OH AFM crystal presents a shortest NO···ON contact at 3.369 Å (the second shortest in the AFM subsets) and the dimers follow an up-down disposition. According to the McConnell-I mechanism all these contacts should present magnetic interactions of the same character (the atoms making the shortest contacts have always the same sign for the atomic spin population). Therefore, such a similarity creates an interesting problem: how two crystals presenting similar ONCNO···ONCNO contacts can have different magnetic interactions? We will see in the next section how this puzzle can be solved.

4
Ab initio Study of the Nature of the Through-Space Magnetic Interactions Present in Nitronyl Nitroxide Crystals: The LICMIT and WILVIW10 Case

The previous sections have shown the weak points of many of the most common magneto-structural correlations used nowadays. Therefore, new and more sound magneto-structural correlations are required to explain the nature of the through-space interactions found in purely organic molecular crystals. These correlations should also serve as guiding principle in the design of new, purely organic molecular crystals.

In this line of thought, we decided to investigate the nature of the magnetic interaction in cases where there are strong conflicts between the McConnell-I predictions and the experimental findings. As described in the previous section, one such conflicting case is found in the LICMIT [27] and WILVIW10 [28] crystals, that is, the α-phase of the 2,5-dihydrophenylnitronyl nitroxide crystal and the *p*-methylpyridylnitronyl nitroxide crystal. The first crystal is the member of the FM subset presenting the shortest N—O···O—N contact (3.158 Å), while the second presents the same property within the AFM subset, being the shortest N—O···O—N distance 3.159 Å. We can assume that the shortest contact is sharing the same magnetic character as the crystal. Therefore, given the fact that the relative disposition of the ONCNO groups in these contacts is similar (Fig. 11), the obvious question is: why two similar ONCNO···ONCNO contacts can have different magnetic nature?

The LICMIT crystal (Fig. 12) orders ferromagnetically below 0.5 K, and shows an effective J/k_B value of +0.93 K [27]. The crystal belongs to the $P2_1/n$ spatial group, with cell parameters $a = 15.142$ Å, $b = 12.320$ Å, $c = 7.196$ Å, and $\beta = 99.18°$. It has four molecules per unit cell, packed forming parallel strips along the c direction, in which the molecules are ordered in an up-up disposition (primary structure) [4d]. The strips form planes (secondary structure) [4d], ordered in a T-shaped disposition among themselves (tertiary structure) [4d]. The shortest NO···ON contacts (Figs. 10 and 11) are found between molecules of parallel strips, connecting radicals in one strip with these in the nearby parallel strip (remember that these molecules present an up-down disposition, previous section).

The WILVIW10 crystal (Fig. 12) shows dominant antiferromagnetic interactions, its effective J/k_B constant being -74 K [28]. It belongs to the P-1 space group, with cell parameters $a = 11.843$ Å, $b = 12.695$ Å, $c = 9.532$ Å, $\alpha = 95.53°$, $\beta = 90.55°$, and $\gamma = 146.89°$. This crystal has two molecules per unit cell, packed forming planes along the b-c directions. Each plane is made by replicating a strip of molecules, ordered within the strip in an up-up disposition. The strips are the primary structure, and the plane is the secondary structure [4d]. These planes then pile up to form the crystal (tertiary structure) [4d]. The shortest NO···ON contacts (Fig. 10) are found within the planes, between adjacent molecules of the same strip.

Given the exponential decay of the strength of the through-space magnetic NO···ON interactions, [29, 30] the dominant magnetic interactions in a crystal

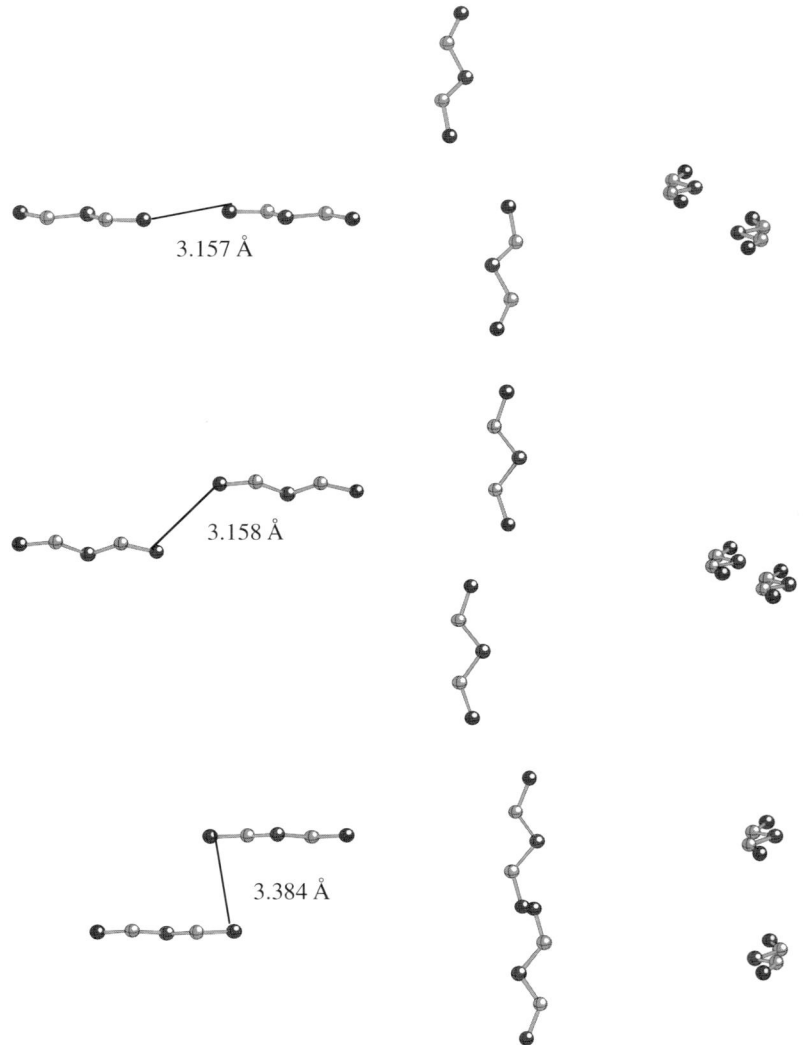

Fig. 11. Frontal, top and side views of the geometry of the shortest ONCNO···ONCNO contacts found in the LICMIT (*upper*) and WILVIW10 (*middle* and *bottom*) crystals. The value of the O···O distance is also indicated

can be expected to be those with the shortest ONCNO···ONCNO contacts, in the case of the LICMIT and WILVIW10 crystals, the contacts shown in Figs. 10 and 11. All the other ONCNO···ONCNO contacts in these two crystals lie at distances higher than 4.3 Å and can be ignored in a first approach.

If one applies the McConnell-I mechanism to the three contacts of Fig. 10, one would expect the three of them to be antiferromagnetic in nature, as the shortest contact is that made by the O atoms of nearby groups (in the 3.384 Å

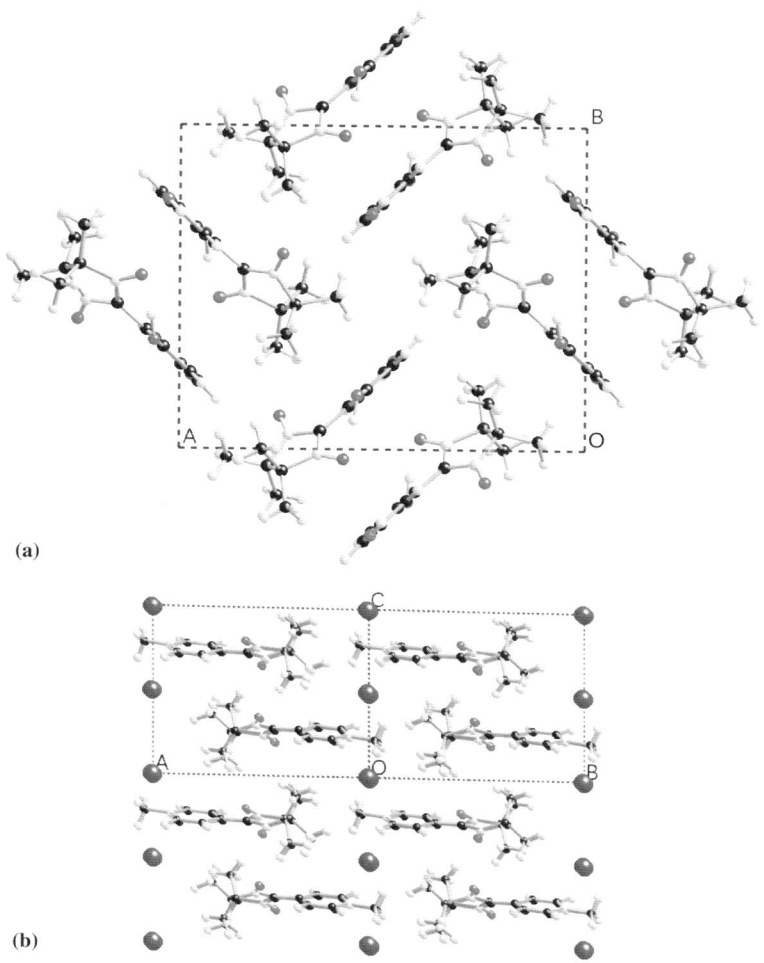

Fig. 12. a Structure of the LICMIT crystal, visualized along the c axis. **b** Structure of the WILVIW10 crystal, visualized along the a axis

contact found in WILVIW10 one can think of an overlap involving also the N atoms, but this does not change the magnetic character of the interaction). Consequently, these contacts involve atoms having the same sign for the atomic spin population. Therefore, the McConnell-I mechanism will predict that these contacts are all antiferromagnetic and, because they are the dominant contacts, will predict these two crystals as presenting dominant antiferromagnetic interactions.

Given the previous incorrect prediction, one has to conclude that the McConnell-I mechanism is oversimplifying the problem along one of the following two possibilities:

(a) the character of the magnetic interaction depends on the small geometrical changes found between the ONCNO···ONCNO contacts of these three isomers (see Fig. 11), or

(b) there are other contacts, besides the ONCNO···ONCNO ones, that are important in defining the magnetic character of these dimers.

We will explore these two possibilities with the help of *ab initio* computations on model systems.

To discard the first possibility, we performed *ab initio* computations on (ONH—C—HNO)$_2$ and (H$_2$NO)$_2$ model dimers, selected to reproduce the interaction of the isolated ONCNO groups of the Figs. 10 and 11. In the (H$_2$NO)$_2$ dimer we explored the magnetic character of the NO···ON contacts when the N—O group is isolated, while in the (ONH—C—HNO)$_2$ dimer we explored the influence on the magnetic character of the NO···ON contacts of the conjugation with the other N—O group of the ONCNO groups. The geometry of the (ONH—C—HNO)$_2$ dimers is that found in the LICMIT and WILVIW10 crystals for the shortest ONCNO···ONCNO groups. It was obtained by deleting all the atoms in Fig. 10 dimers except those in the two ONCNO groups, and then attaching an H atom to each NO group at the same distance it has in the H$_2$NO molecule. The (H$_2$NO)$_2$ model dimer was obtained by also deleting the C and NO atoms not belonging to the NO group making the shortest contact. The magnetic character of the interaction in these model dimers can be obtained by computing the singlet-triplet energy difference in the dimer [$\Delta E^{S-T} = E(S) - E(T)$]. A positive value indicates a triplet ground state, that is, a ferromagnetic interaction, while a negative value is obtained when the singlet is the ground state and the interaction is antiferromagnetic.

Previous computations have shown that one can obtain qualitative values of the singlet-triplet energy differences at the CASSCF level, [29, 31] although a precise computation of these values requires methods which include the dynamical electron correlation, like the CASPT2 methods. Alternatively, one can obtain quantitative the singlet-triplet energy differences in these dimers using dedicated energy difference methods [31b]. Table 4 collects the values of the ΔE^{S-T} values obtained for the two model dimers of the three dimers of Fig. 10. The CASSCF computations on the (H$_2$NO)$_2$ model dimer were done using a (2,2) complete active space, which produces the same qualitative singlet-triplet energy differences as the larger spaces used in Sect. 3.2. However, a larger (6,6) active space was used for the (ONH—C—HNO)$_2$ model dimer, in order to describe the polarization effects found in each monomer. These effects are responsible for the presence of negative atomic spins in the central C atom of these monomers. The computations on the smaller dimer were done with three basis sets, the 6-31G, 6-31G(d), and 6-31+G(d) basis, to explore the basis set truncation effects. On the larger dimer, we used only the smaller two basis sets for the same evaluation, since the addition of diffuse functions did not strongly affect the ΔE^{S-T} results in the (H$_2$NO)$_2$ dimer.

A look at the results in Table 4 shows that, independently of the basis set and model dimer used, the singlet is the ground state in the two dimers, that is,

Table 4. Values of the ΔE^{S-T} energy difference computed for the (ONH—C—HNO)$_2$ and (H$_2$NO)$_2$ model dimers using the CASSCF and B3LYP *broken symmetry* methods (see text)

Model system	Crystal	Basis set	ΔE^{S-T} CASSCF cm^{-1}	ΔE^{S-T} B3LYP cm^{-1}
(H$_2$NO)$_2$	LICMIT	6-31G	−3.4	−4.6
		6-31G(d)	−4.2	−4.4
		6-31G + (d)	−4.2	−8.0
	WILVIW10	6-31G		
		6-31G(d)		
		6-31 + G(d)		
(ONH—C—HNO)$_2$	LICMIT	6-31G	−0.8	−2.0
		6-31G(d)	−1.0	−1.8
	WILVIW10	6-31G	−2.2^a/−16.4^b	−5.4^a/−14.0^b
		6-31G(d)	−1.8^a/−17.6^b	−5.2^a/−15.0^b

^{a} Dimer with a NO···ON contact of 3.159 Å in Fig. 7;
^{b} Dimer with a NO···ON contact of 3.383 Å in Fig. 7.

the magnetic interaction in these dimers is always antiferromagnetic. The numbers also show a non-negligible influence of the second group in the singlet-triplet stability, which becomes a lot less antiferromagnetic in the larger dimer. A CASPT2 calculation on the CASSCF(6,6) wave function of the (ONH—C—HNO)$_2$ model dimer confirmed the higher stability of the singlet state (by 2.28 cm^{-1} in the LICMIT geometry, and by 0.77 and 9.68 cm^{-1} in the WILVIW10 geometries). The CASPT2 values can be taken as a reference for other less accurate methods than the CASSCF (in this case, one finds that the CASSCF singlet-triplet separations have the right sign but are smaller than the CASPT2 numbers, although of the same order of magnitude). In conclusion, the results from these highly accurate *ab initio* methods confirm the McConnell-I prediction and indicate that looking only at the ONCNO···ONCNO groups it is not possible to reproduce the magnetic character of the dimers found in the LICMIT crystal. This fact justifies the lack of correlation between the magnetic character of the interactions and the relative geometry of the ONCNO groups reported in the previous statistical analysis of the nitronyl nitroxide crystals.

Given the impossibility to reproduce the experimental character of the magnetic interaction by looking only at the ONCNO groups, one can suspect that the nature of the magnetic interaction is strongly affected by atoms whose atomic spin density is known to be small. To test at the *ab initio* level the validity of such a proposition, we computed the singlet-triplet energy difference (ΔE^{S-T}) using the whole geometry of the dimers of Fig. 10. Given the size of these molecules, the computations were done using a broken-symmetry formulation [32] of the unrestricted B3LYP determinant (hereafter identified as BS-UB3LYP), where by B3LYP we mean the usual three parameters non-local exchange and correlation functional [33]. Such a formulation has been shown to provide a very precise description of the singlet-triplet gaps in organometallic complexes [34]. We have tested the quality of the BS-UB3LYP computations on our dimers by recomputing the singlet-triplet gap of Table 4 for the (ONH—C—HNO)$_2$ and (H$_2$NO)$_2$

dimers (last column in Table 4). The results show that the BS-UB3LYP singlet-triplet splitting has always the same sign and order of magnitude as the splitting computed at the CASSCF level, although the BS-UB3LYP values are always a bit larger (in fact, close to the CASPT2 values). Therefore, we can feel confident about the ability of the BS-UB3LYP method to compute the nature of the through-space magnetic interactions.

The BS-UB3LYP computations on the whole LICMIT dimer gave a ΔE^{S-T} value of 1.3 cm^{-1} [34] (to be compared with an experimental value of 1.29 cm^{-1}), while for the two WILVIW10 dimers the values were −9.1 cm^{-1} for the shortest one, and −28.2 cm^{-1}, for the largest one (to be compared with an experimental averaged value of −102.9 cm^{-1}) [34]. Therefore, when the whole geometry of the dimer is included, the *ab initio* methods can reproduce the experimental character of the magnetic interaction in the LICMIT and WILVIW10 crystals. This is a numerical proof that *the atoms of the radical which hold in total only 10% of the atomic spin density play an important role in determining the nature of the magnetic interaction*. An inspection of the geometry of the LICMIT dimer and the WILVIW10 dimers suggests that the NO···aromatic ring contacts could be the key interaction in establishing the ferromagnetism in the LICMIT dimer; these are left out when only the ONCNO groups are considered. However, more detailed studies are needed before this fact can be stated without discussion. In any case, the *ab initio* computations on the whole dimer have shown that the magnetic character of the interaction is not a consequence of the relative geometry of only one functional group, but depends on the relative positions of all the groups. This relative position of many functional groups receives the name *pattern*. Therefore, we can talk about *magnetic patterns*, that is, the patterns found in magnetic crystals, each of them having a magnetic net interaction, not necessarily of the same nature in all the patterns of the same crystal. The nature of the pattern is not solely defined by the relative geometry of the ONCNO groups. Consequently, although the orientations of these groups in various patterns can be similar, the magnetic character of patterns sharing the same ONCNO relative orientations is not necessarily the same. This explains the apparent inconsistency found in the statistical analysis of the nitronyl nitroxide crystals presenting dominant magnetic interactions, in particular, the lack of dependence of the magnetic character with the relative orientation of the ONCNO groups.

To complete our evaluation of the character of the magnetic interactions present in the LICMIT crystal, the only point that remains to be tested is our hypotheses that the dominant magnetic character of the crystal was associated to the shortest contact. Therefore, we computed the singlet-triplet energy difference for *all* the nearest neighbors that one of the radicals makes with the surrounding ones, irrespective of the type of contacts found between these radicals and their shortest distance. We found [35] five different types of nearest neighbor dimers in the LICMIT crystal, plotted in Fig. 13. They present NO···ON distances of 3.158, 4.594, 5.525, 5.856, and 6.294 Å, the shortest one being that for the dimer of Fig. 10. The singlet-triplet energy difference in each of these dimers are 1.3, 0.15, 0.04, −0.15, and 0.11 cm^{-1}, respectively [35].

Therefore, the strongest magnetic interaction is by far the shortest one (by one order of magnitude) and will dominate over the rest of radical···radical interactions. It is interesting to note the ferromagnetic character of all but one of the nearest neighbors radical···radical interactions in the LICMIT crystal. The same applies to the WILVIW10 crystal [35]. Consequently, the dominant magnetic assumption made in the analysis of the LICMIT and WILVIW10 crystals seems to be a valid one.

5
Concluding Remarks

From the previous studies we can extract some useful conclusions for a better understanding of the through-space magnetism found in purely organic molecular crystals. First of all, we have seen that the McConnell-I mechanism, the most popular used nowadays, lacks of a consistent methodological basis and part of its success in the [2,2]paracyclophanes relies in a fortuitous error cancellation due to the high symmetry of these biradicals. The McConnell-I predictions are shown to fail in the $H_2NO···ONH_2$ dimers, and also in some orientations of the methyl···allyl dimer. We have also presented one example of a nitronyl nitroxide crystal in which McConnell-I fails, and discussed the problems found when using this mechanism on nitronyl nitroxides presenting packings driven by hydrogen bonds.

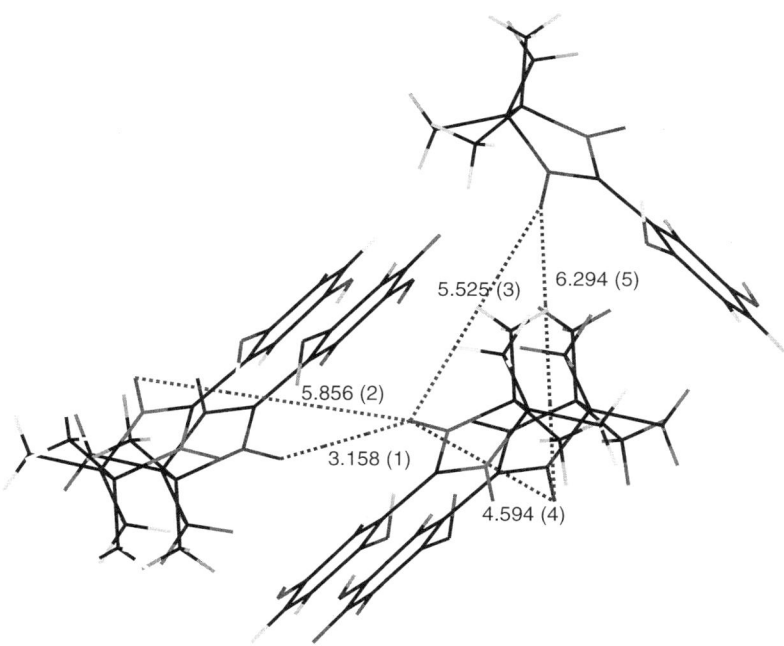

Fig. 13. Different types of nearest neighbor dimers found around a radical in the LICMIT crystal. The O···O distance is given for each dimer

A detailed statistical analysis of the packing of many nitronyl nitroxide crystals presenting dominant magnetic interactions was also done. No preliminary assumptions were made on which geometrical parameters were going to play a determinant role in defining the magnetic character. This analysis shows that the magnetic character of the crystals does not present any relationship with the relative orientation of the *individual* ONCNO···ONCNO or C—H···ONCNO contacts. Therefore, it suggested that some kind of cooperative effect among all the functional groups present in the molecule is at work in these crystals. *Ab initio* studies on the LICMIT and WILVIW10 crystals provides a numerical proof of existence on these types of cooperative effect: while one cannot explain the ferromagnetic character of the LICMIT crystal looking only at the relative orientations of the ONCNO groups, the ferromagnetism is found when the six-membered rings of the radicals are included in the computation. This proves that the character of these magnetic interactions in the dimer is determined by *all* groups present in the dimer, regardless of the atomic spin population of their constituting atoms. That is, we have to look at the *magnetic patterns* (combination of short contacts defining a energetically stable supramolecular aggregate of the magnetic crystal). Each presents a dominant magnetic interaction, resulting from the combination of the magnetic interactions generated by the individual contacts, which depend on all the functional groups present in the pattern. In the LICMIT dimer, it seems that the NO···aromatic ring contact works in conjunction with the ONCNO···ONCNO contacts in establishing the magnetic character of the dimer, but further numerical tests are needed to fully prove this point.

As a consequence of that cooperativity, one cannot expect to carry out reliable magneto-structural analysis of purely organic molecular crystals using oversimplified empirical magneto-structural correlations which assume that only one group is determining the magnetic character of the dimer. This fact discards the oversimplified analysis performed using the McConnell-I mechanism in which the position of the ONCNO groups are analyzed. We have also seen that McConnell-I also fails when other groups are considered, in part because of the simplistic form in which it is generally used (the exchange integrals are not considered). Further studies are under way to be able to define reliable magneto-structural correlations based on the correct assumptions. At the present moment, these studies [35] seem to indicate that one should use a more elaborate mechanism, like some sort of improved version of the McConnell-II mechanism, capable of solving the problems detected by Kahn and others in some well known magnets, but along the avenues described by Kinoshita in the first chapter of this book [36].

Acknowledgements. This work was supported by DGES (Projects PB95-0848-C02-02 and PB98-1166-C02-02) and CIRIT (Projects 1997SGR-00072 and 1999SGR-00046), and by the generous allocation of CPU time in the computers of CESCA and CEPBA made possible by a joint CIRIT-University of Barcelona grant program. The authors also want to thank Prof. J. Veciana and M. A. Robb for their invaluable support and cooperation in parts of this work.

6
References

1. Recent reviews of the field can be found in: (a) Miller JS, Epstein AJ (1994) Angew Chem Int Ed Engl 33: 385; (b) Kinoshita M (1994) Jpn J Appl Phys 33: 5718; (c) Kahn O (1993) Molecular Magnetism. VCH, New York; (d) Lahti P (ed) (1999) Magnetic Properties of Organic Materials. Marcel Dekker, New York; (e) Gatteschi D, Kanh O, Miller JS, Palacio F (eds) (1991) Molecular Magnetic Materials. Kluwer Academic, Dordrecht; (f) Coronado E, Delhaes P, Gatteschi D, Miller JS (eds) (1996) Molecular Magnetism: From Molecular Assemblies to Devices. Kluwer Academic, Dordrecht; (g) Kahn O (ed) (1996) Magnetism: A Supramolecular Function. Kluwer Academic, Dordrecht; (h) Turnbull MM, Sugimoto T, Thompson LK (eds) (1996) Design of Organic-Based Materials with Controlled Magnetic Properties. ACS Symposium Series 664, Washington; (i) All the Proceedings of the International Conference on Molecule Based Magnets, the latest being: Kahn O (ed) (1999) Mol Cryst Liq Cryst 334–335: 1–712, 1–706
2. (a) Anderson PW (1963) In: Rado GT, Shul H (eds), Magnetism. Academic Press, New York, Vol 1, p 25; (b) Hay PJ, Thibeault JC, Hoffmann R (1975) J Am Chem Soc 97: 4884; (c) Kollmar C, Kahn O (1993) Acc Chem Res 26: 259
3. See, for instance, the case of the *para*-nitrophenylnitronyl nitroxide polymorphs recently reviewed in: Kinoshita M (1997) In: Nalwa SH (ed), Handbook of Organic Molecules and Polymers. John Wiley, New York, Vol 1, p 781
4. Reviews on crystal packing analysis can be found in: (a) Gavezzotti A (1998) Cryst Rev 7: 5; (b) Gavezzotti A (1994) Acc Chem Res 27: 309; (c) Desiraju GR (1995) Angew Chem Int Ed Eng 34: 2311; (d) Novoa JJ (1999) In: Howard JAK, Allen FH, Shields GP (eds), Implications of Molecular and Materials Structure for New Technologies. Kluwer, Dordrecht
5. An study of the crystal packing and polymorphism of the simplest nitronyl nitroxide radical, the hydro derivative, can be found in: (a) Filippini G, Gavezzotti A, Novoa JJ (1999) Acta Cryst B55: 543; (b) Deumal M, Cirujeda J, Veciana J, Kinoshita M, Hosokoshi Y, Novoa JJ (1997) Chem Phys Lett 265: 190
6. Desiraju G (1989) Crystal Engineering. The Design of Organic Solids. Elsevier, Amsterdam
7. See, for instance: Lafranc de Panthou F, Luneau D, Laugier J, Rey P (1993) J Am Chem Soc 115: 9100
8. McConnell HM (1963) J Chem Phys 39: 1910
9. The second mechanism is the charge-transfer mechanism published in: McConnell, HM (1967) Proc Robert A Welch Found Conf Chem Res 11: 144. A historical recount is given by this author in ref. [1d]
10. Izuoka A, Murata S, Sugawara T, Iwamura H (1987) J Am Chem Soc 109: 2631
11. Deumal M, Novoa JJ, Bearpark MJ, Celani P, Olivucci M, Robb MA (1998) J Phys Chem A 102: 8404
12. For a review see: van Vleck JH (1945) Rev Mod Phys 17: 27
13. (a) Bernardi F, Olivucci M, McDouall JJ, Robb MA (1988) J Chem Phys 89: 6365; (b) Bearpark MJ, Bernardi F, Olivucci M, Robb MA (1994) Chem Phys Lett 217: 513; (c) Bernardi F, Olivucci M, Robb MA (1992) J Am Chem Soc 114: 1606
14. Awaga K, Maruyama Y (1989) J Chem Phys 91: 2743
15. Zheludev A, Barone V, Bonnet M, Delley B, Grand A, Ressouche E, Rey P, Subra R, Schweizer J (1994) J Am Chem Soc 116: 2019
16. A general reference on density functional theory is: Parr RG, Yang W (1989) Density Functional Theory of Atoms and Molecules. Oxford University Press, New York
17. BLYP stands for a combination of the Becke functional for the non-local exchange, and the Lee-Yang-Parr functional for the non-local exchange. See: (a) Becke AD (1988) Phys Rev A 38: 3098; (b) Lee C, Yang W, Parr RG (1988) Phys Rev B 37: 785

18. Ditchfield R, Hehre WJ, Pople JA (1971) J Chem Phys 54: 724
19. (a) Heise H, Köhler FH, Mota F, Novoa JJ, Veciana J (1999) J Am Chem Soc 121: 9659; (b) Cirujeda J, Vidal-Gancedo J, Jürgens O, Mota F, Novoa JJ, Rovira C, Veciana J (2000) J Am Chem Soc 122: 11393
20. For a rigorous definition of the atomic volume see: Bader RWF (1990) Atoms in Molecules. A Quantum Theory. Oxford University Press, Oxford. Numerical tests have shown that on the case of the nitronyl nitroxides the atomic spin population obtained by rigorous integration of the atomic spin volume is very similar to that obtained using the Mulliken Population Analysis: Novoa JJ, Mota F, Veciana J, Cirujeda J (1995) Mol Cryst Liq Cryst 271: 79
21. (a) Veciana J, Cirujeda J, Rovira C, Vidal-Gancedo J (1994) Adv Mater 4: 1377; (b) Akita T, Mazaki Y, Kobayashi K (1995) J Chem Soc Chem Commun 1861; (c) Veciana J, Cirujeda J, Rovira C, Vidal-Gancedo J (1995) Adv Mater 2: 221
22. Kawakami T, Takeda S, Mori W, Yamaguchi K (1996) Chem Phys Lett 261: 129
23. Novoa JJ, Deumal M, Lafuente P, Robb MA (1999) Mol Cryst Liq Cryst 335: 603
24. Deumal M, Cirujeda J, Veciana J, Novoa JJ (1999) Chem Eur J 5: 1631
25. A list of the crystals in each subset, including their refcodes (when available) are given in ref. [23]
26. Cirujeda J (1997) Ph. D. Thesis, Ramon Llull University
27. LICMIT: (a) Sugawara T, Matsushita MM, Izuoka A, Wada N, Takeda N, Ishiwara M (1994) J Chem Soc Chem Commun 1723; (b) Matsushita MM, Izuoka A, Sugawara T, Kobayashi T, Wada N, Takeda N, Ishiwara M (1997) J Am Chem Soc 119: 4369
28. WILVIW10: Awaga K, Yamaguchi A, Okuno T, Inabe T, Nakamura T, Matsumoto M, Maruyama Y (1994) J Mater Chem 4: 1377
29. Yamanaka S, Okumura K, Yamaguchi K, Hirao K (1994) Chem Phys Lett 225: 213
30. We have also seen an exponential decay in the H_2NO dimers oriented as in Fig. 5, for the 180° and 90° conformations. The computations were done using the CASSCF method, a (6,4) active space, and the 6-31 + G(d) basis set
31. (a) Bauschlicher CW, Langhoff SR, Taylor PR (1989) Adv Chem Phys 77: 103; (b) Castell O, Caballol R, Subra R, Grand A (1995) J Phys Chem 99: 154
32. (a) Noodleman L (1981) J Chem Phys 74: 5737; (b) Noodleman L, Davidson ER (1986) Chem Phys 109: 131; (c) Hart JR, Rappe AK, Gorum SM, Upton TH (1992) J Phys Chem 96: 6264; (d) Caballol R, Castell O, Illas F, Moreira I de PR, Malrieu JP (1997) J Phys Chem A 101: 7860
33. The B3LYP is an exchange-correlation hybrid functional which includes a mixture of Hartree-Fock and Becke exchange with VWN and LYP correlation functional: (a) Becke AD (1988) Phys Rev A 38: 3098; (b) Lee C, Yang W, Parr RG (1993) Phys Rev B 37: 785
34. See, for instance: Ruiz E, Alemany P, Alvarez S, Cano J (1997) J Am Chem Soc 119: 1297
35. Deumal M, Novoa JJ (submitted for publication)
36. Kinoshita, M (2001) chapter 1 of this book

Metal-Aminoxyl-Based Molecular Magnets

Katsuya Inoue

Institute for Molecular Science, 38 Nishigounaka, Myodaiji, Okazaki, 444-8585, Japan
E-mail: kino@ims.ac.jp

The nitroxide radical is one example of a stable organic radical, and the oxygen atom of the nitroxide group exhibits weak basicity; therefore, it demonstrates coordinating ability with respect to transition metal ions. There are many π-conjugated oligo-nitroxides which have high-spin ground states. When the oligo-nitroxides are used as a bridging ligand for paramagnetic metal ions, we can control both of the structure of the complex and the spin structure. This chapter discusses one-dimensional, two-dimensional, and three-dimensional magnetic materials derived from coordination complexes consisting of magnetic metal ions and nitroxide radicals.

Keywords: Organic radicals, Magnetism, Transition metal complexes

1	Introduction	62
2	Nitroxide Radicals	62
2.1	Magnetic Interaction Between the Transition Metal Ions and Nitroxide Radicals Through Coordination Bonds	64
2.2	Design Strategy for Nitroxide Radical-Metal Magnetic Complexes	68
2.3	Preparation of Metal-Poly(Nitroxide) Radical Complexes	69
3	One-Dimensional Nitroxide-Metal Systems	73
3.1	Structure and Magnetic Properties of Ferrimagnetic 1-D Chain Formed by Manganese(II) and Nitroxide Radicals	75
3.1.1	Manganese(II) Complexes with Nitronyl Nitroxides	75
3.1.2	Manganese(II) Complexes with Triplet Bis-Nitroxide Radicals . .	75
3.2	Conclusion	77
4	Two-Dimensional Metal-Nitroxide Systems	78
4.1	Structure and Magnetic Properties of Ferrimagnetic 2-D Sheet Formed by Manganese(II) and Nitroxide Radicals	80
4.1.1	2-D Manganese(II) Complexes with Nitronyl Nitroxides	80
4.1.2	Manganese(II) Complexes with Triplet Bis-Nitroxide Radicals . .	81

Dedicated to the memory of Prof. Olivier Kahn

4.2	Conclusion	84
5	**Three-Dimensional Metal-Nitroxide Systems**	86
5.1	Crystal and Molecular Structures	86
5.2	Magnetic Properties of 3-D Systems	86
5.3	Conclusion	88
6	**Conclusions**	89
7	**References**	89

1
Introduction

The construction of molecular-based magnets that have a well-defined magnetic structure is a scientific topic of increasing interest. The idea is to establish unprecedented macroscopic spins of long-range order in molecular systems. There are mainly three steps to generate molecule-based magnets:

i) align the magnetic moments in the molecule;
ii) arrange the molecules in the crystals; and
iii) order all the spins in a mesoscopic scale.

In approaches employing purely organic crystals there are many difficulties in steps ii) and iii). A network structure may not be a problem if an appropriate crystal design is made. However, since the exchange coupling between neighboring molecules through van der Waals forces, hydrogen bonding, hydrophobic interaction, etc., is not necessarily strong, it is difficult to expect strong intermolecular magnetic coupling. There are principally two conditions to obtain high-T_C magnets:

1) make strong exchange interactions between spins and
2) all the spins are ordered in two- or three-dimensions.

In order to satisfy the above conditions, it is necessary to create polymeric networks among spins. It is, therefore, a powerful design strategy to make extended polymeric structures by assembling free radicals by means of magnetic metal ions. Such polymeric complexes satisfy the above-mentioned conditions (i–iii). In this chapter magnetic materials made up of coordination complexes of magnetic metal ions and nitroxide radicals are discussed.

2
Nitroxide Radicals

In order to use free radicals as ligands for a magnetic metal, it is necessary for the radicals to have sufficient coordinating ability [4–9]. The oxygen atom of the nitroxide group has a weak basicity; and thus has coordinating ability to transition metal ions.

There are some π-conjugated oligoradicals that have a high-spin ground state. Various bis-nitroxide radicals with triplet ground states and tris-nitroxide radicals with quartet ground states have been designed on the basis of the *m*-phenylene topology and prepared by oxidation of the corresponding hydroxyamines. The magnetic interaction between nitroxides depends on the distances. High-spin oligo-nitroxide radicals that have been used as ligands for magnetic metal ions are summarized in Scheme 1 [10–18].

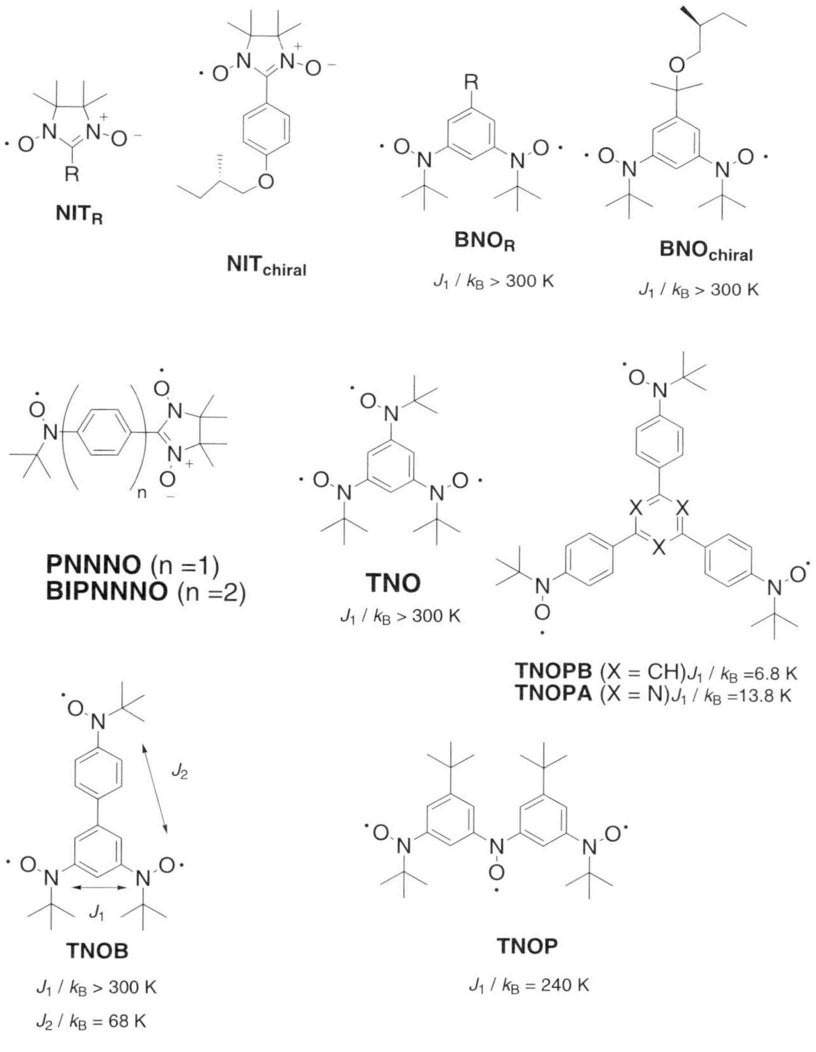

NIT$_R$

NIT$_{chiral}$

BNO$_R$
$J_1 / k_B > 300$ K

BNO$_{chiral}$
$J_1 / k_B > 300$ K

PNNNO (n = 1)
BIPNNNO (n = 2)

TNO
$J_1 / k_B > 300$ K

TNOPB (X = CH) $J_1 / k_B = 6.8$ K
TNOPA (X = N) $J_1 / k_B = 13.8$ K

TNOB
$J_1 / k_B > 300$ K
$J_2 / k_B = 68$ K

TNOP
$J_1 / k_B = 240$ K

Scheme 1.

2.1
Magnetic Interaction Between Transition Metal Ions and Nitroxide Radicals Through Coordination Bonds

Syntheses of metal-nitroxide complexes were first reported in the early 1970s as spin labels for metal complexes or catalysts for living polymerizations. In principle, when the overlap between the magnetic orbitals is large, a normal covalent bond will be formed and the spins are paired. In practice, however, the bond is only partial as judged from the observed pairing energy of the electrons. On the other hand, when the overlap is small or zero, direct exchange will be operative, giving either weak antiferro- or ferromagnetic coupling. The coupling modes are classified in terms of the geometry of the two magnetic orbitals involved.

Structures of metal-nitroxide complexes are classified in the first approximation into two limiting arrangements, the nitroxide oxygen occupying an axial or an equatorial coordination site. In both limits, the magnetic orbital of copper(II) may be loosely described as xy, lying in the equatorial plane of the pyramid with the lobes pointing towards the ligands. When the nitroxide ligand occupies an axial position, its π^* magnetic orbital is essentially orthogonal to the magnetic orbital of copper(II) and the coupling is expected to be ferromagnetic. Indeed, this has been found to be the case for a number of complexes in which magnetic susceptibility data showed the existence of ferromagnetic coupling in the range 30–100 K [19–23]. On the other hand, when the nitroxide oxygen atom is in the equatorial (or basal) position in copper(II) complexes, the exchange interaction tends to be antiferromagnetic [19–21].

In the case of high spin d^5 manganese(II), at least one of five magnetic orbitals must have the correct symmetry for overlapping with those of the nitroxides. As a result, the interaction is antiferromagnetic and a ground $S = 3/2$ (=5/2 − 1/2 − 1/2) state will result in complexes of the general formula [Mn(hfac)$_2$(nitroxide)$_2$]. For TEMPO and PROXY ligands, J/k_B values are −227 and −302 K, respectively [19–21, 24, 25].

3-Imidazoline nitroxides and Ullman's nitronyl nitroxides display have basic centers and, therefore, serve as good bridging ligands for extended structures with coordinatively doubly unsaturated metal ions. These species will be discussed in Sect. 3 and 4. At the same time, since Cu(II) has the ability to form complexes with coordination numbers of both 5 and 6, trinuclear complexes with a 3:2 stoichiometry [3Cu(hfac)$_2$ · 2nitroxide] are obtained [26, 27]. The magnetic coupling in these complexes is ferromagnetic except for one with the 2-ethyl derivative of nitronyl nitroxide [26]. In general, exchange coupling of Mn(II), Ni(II), and Co(II) with nitronyl nitroxide ligand is antiferromagnetic in character.

There are cyclic transition metal complexes with nitroxide radicals or nitronyl nitroxide radicals. The number of spins is finite for these complexes; therefore, the systems can afford a rigorous solution of the spin Hamiltonian to give the sign and magnitude of the exchange coupling. These values serve as good measures for designing and analyzing of new extended systems containing two or more types of exchange coupling parameters.

The cyclic dimer complex of bis(hexafluoroacetylacetonato)manganese(II) with 5-*tert*-butyl-1,3-phenylenebis(*N*-*tert*-butylnitroxide) [28, 29] is such an example. The systems have been analyzed in order to obtain two sets of two intramolecular exchange coupling parameters J_1 and J_2 (Scheme 2). The crystal structure (Fig. 1) of the first complex, [Mn(hfac)$_2$ · **BNO**$_{t\text{-Bu}}$]$_2$, is found to be a cyclic analog of 1-D complex polymers made of similar components lacking a ring system, a *tert*-butyl moiety, or with halogen substituents in its place (see Sect. 3) [30–33]. The molecule consists of two Mn(II) ions ($S = 5/2$) and four nitroxide radicals ($S = 1/2$) arranged in a cyclic array as shown in Scheme 2. The molar paramagnetic susceptibility χ_{mol} of the complex was investigated in the temperature range 2–300 K. The $\chi_{mol}T$ values exhibit a continuous increase from room temperature to ca 50 K, reaching a plateau at 5.71 K cm^3 mol^{-1}.

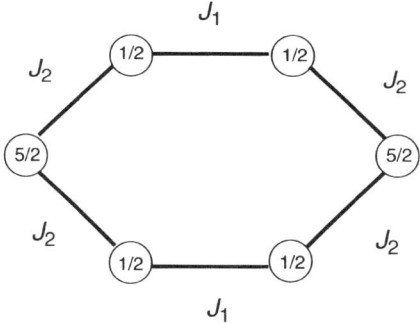

Scheme 2.

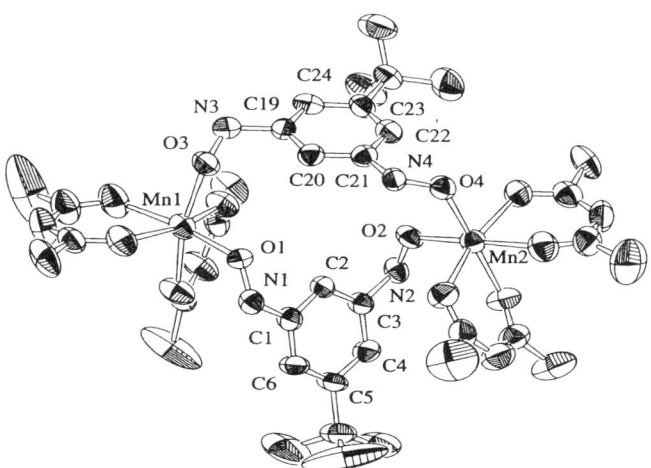

Fig. 1. Cluster structure of [Mn(hfac)$_2$ · **BNO**$_{t\text{-Bu}}$]$_2$. Fluorine atoms, *N*-*tert*-butyl groups and hydrogen atoms are omitted. Thermal ellipsoids are drawn at the 50% probability level

This agrees well with the theoretical spin-only value for an $S = 3$ spin ground state, which can be explained by ferromagnetic (F) interaction between two $S = 3/2$ pseudo-spins ($5/2 - 2 \times 1/2$), obtained by assuming a predominant antiferromagnetic (AF) coupling (J_2) between the manganese(II) ions and the two coordinated radicals. This high-spin ground state is stabilized with respect to upper energy levels by several hundred wave numbers, as deduced from the decrease in $\chi_{mol}T$ with increasing temperature. A decrease in $\chi_{mol}T$ at the lowest temperatures may be analyzed in terms of intermolecular AF coupling.

The $\chi_{mol}T$ vs. T plots were analyzed in two ways: one is a rigorous solution of the spin Hamiltonian for the six-spin system [34–36], and the other by a model of the coupling of a cyclic array of effective spins $S = 3/2$. The first method together with the molecular field contribution taken into account by the Curie–Weiss law gives $J_1/k_B = +1005$ K (± 11 K), $J_2/k_B = -539$ K (± 19 K) and Weiss constant $\lambda = -0.10$ K (± 0.01 K), with $g = 1.953$ [37], provides a very good description of the experimental data over the entire temperature range. The second method gave $J/k_B = +68.6$ K (± 0.6 K) for repeating $S = 3/2$ units and the Lande factor $g = 2.002$. Simulation of the observed data is quite good except that, the high-temperature data are poorly reproduced by the second model. This is due to the lack of precision in the approximation method in locating correctly the higher energy levels where spins are more populated at higher temperature.

A zero-dimensional complex of bis(hexafluoroacetylacetonato)manganese(II) and 4-N-*tert*-butylnitroxide-4'-(1-oxyl-3-oxide-4,4,5,5-tetramethylimidazolin-2-yl)biphenyl (**BIPNNNO**) [38] is also such an example. The complex consists of discrete molecules of [Mn(hfac)$_2$] · (**BIPNNNO**)$_2$ and has a triclinic structure (space group $P\bar{1}$) where $a = 11.267(3)$ Å, $b = 12.605(2)$ Å, $c = 10.661(3)$ Å, $\alpha = 99.16(1)°$, $\beta = 96.97(3)°$, $\gamma = 82.60(1)°$, $V = 1471.5(6)$ Å^3. The [Mn(hfac)$_2$] · (**BIPNNNO**)$_2$ molecule is shown in Fig. 2. The Mn(II) ion lies on a crystallographic inversion center and is coordinated by a pair of **BIPNNNO** moieties and a pair of bidentate hfac anions, which afford an MnO$_6$ octahedral environment. The two **BIPNNNO** molecules coordinate *trans* to one another. The Mn—O bond distances are 2.128(5) Å for the **BIPNNNO** radical and 2.153(5) and 2.130(5) Å for the hfac ligands. The formation of a 1:2 complex, [Mn(hfac)$_2$] · (**BIPNNNO**)$_2$, suggests the higher coordination ability for N-*tert*-butyl nitroxide than for nitronyl nitroxide to the Lewis acidic manganese ion. The temperature dependence of $\chi_{mol}T$ of the complex is illustrated in Fig. 3. The low temperature value of $\chi_{mol}T$ tends toward 0.375 emu/mol, a limit for the $S = 1/2$ ground state. This value is expected for the negative Mn—NO and positive intraradical NO—NN exchange parameters. The complex remains paramagnetic down to 1.8 K. As demonstrated in Fig. 3, the magnetization curve at this temperature can be well described by a Brillouin function for $S = 1/2$, $B_{1/2}(H/T)$, neglecting the intercluster exchange interaction.

The high temperature limit expected for separately-acting four spins 1/2 and one spin 5/2, 5.875 emu/mol, is far from being reached. The character of $\chi_{mol}T$ temperature dependence indicates that both of the exchange interactions are comparable and that new levels are successively excited with increasing

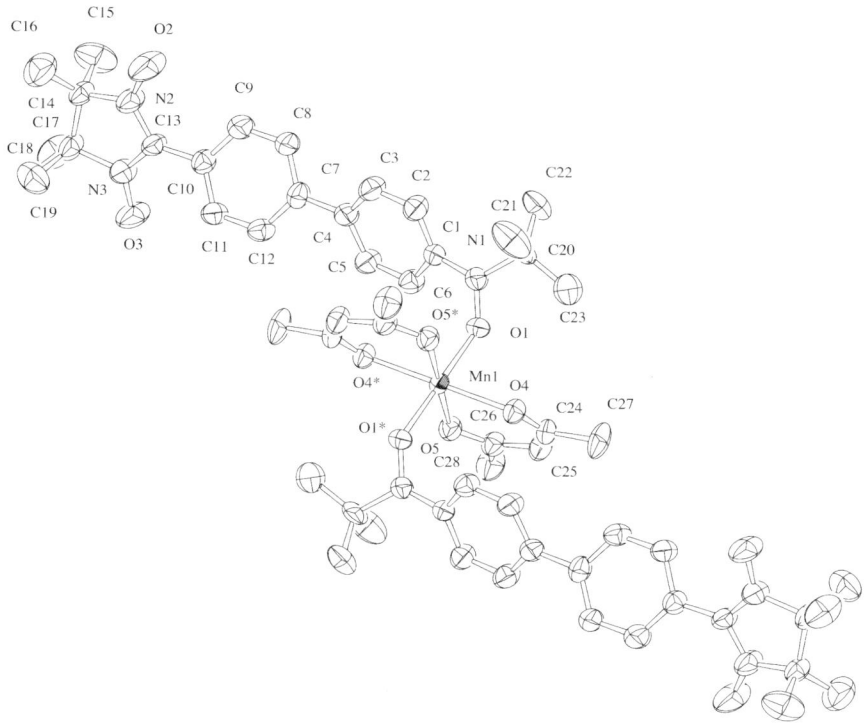

Fig. 2. View of the Mn(hfac)$_2$(**BIPNNNO**)$_2$ molecule

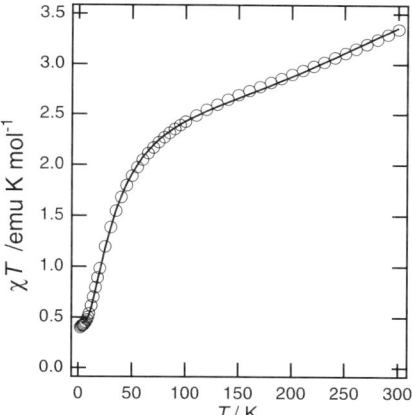

Fig. 3. Temperature dependence of $\chi_{mol}T$ of the complex Mn(hfac)$_2$(**BIPNNNO**)$_2$. Experimental data are denoted by symbols. Solid line corresponds to the theoretical fit (see text)

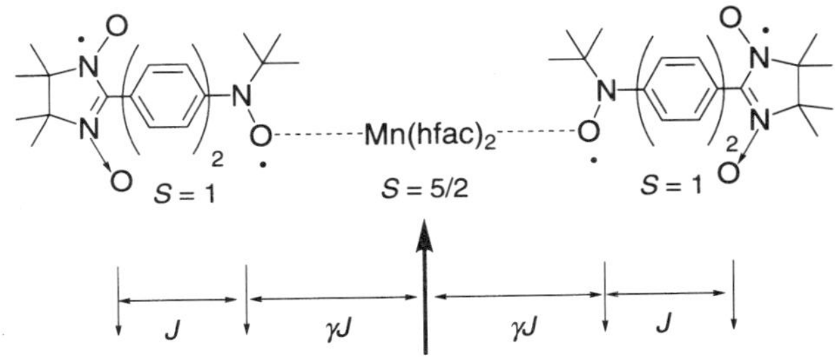

Fig. 4. Magnetic structure of Mn(hfac)$_2$(BIPNNNO)$_2$

temperature T up to room temperature. The temperature variation of $\chi_{mol}T$ was simulated numerically in the Heisenberg approximation with use of the Van Vleck equation. The energy level scheme was determined by exact diagonalization of the following Hamiltonian (Fig. 4):

$$H = -2J[S_1S_2 + \gamma(S_2S_3 + S_3S_4) + S_4S_5] \tag{1}$$

where J and γJ are the NO—NN and Mn—NO exchange integrals, respectively, and $\hat{S}_3$ is the Mn^{2+} spin operator. The best fit parameters are $J/k_B = 72 \pm 5$ K and $\gamma = -1.9$ ($\gamma J/k_B = -135 \pm 10$ K) with the purity factor $P = 1.1$.

2.2
Design Strategy for Nitroxide Radical-Metal Magnetic Complexes

As we have previously seen, a set of two conditions makes the necessary and sufficient conditions for realizing metal-radical complexes having interesting magnetic properties.

1) A ligand must possess at least two nitroxide coordination sites. Any nitroxide group may form a metal complex having this nitroxide as a ligand, but there is no chance of forming an extended structure without additional coordination sites.
2) When high-spin nitroxide radicals are utilized as ligands for the complex, the interaction among nitroxides must be ferromagnetic.

These are the reasons why nitronyl nitroxides and high-spin oligo-nitroxide radicals are of importance for material design. When an organic free radical ligand has two ligation sites, e.g., nitronyl nitroxides or bis-nitroxide radicals, extended complex structures that are often chain polymers or macrocycles, are formed with coordinatively doubly unsaturated metal ions. Extension of this design strategy to tris-nitroxide radicals exhibiting quartet ground states has led to the construction of two- (2-D) and three-dimensional (3-D) network structures in which both organic $2p$ and metallic $3d$ spins have been ordered in macroscopic scales [1, 3, 39–41]. This schematic is shown in Fig. 5.

π-Conjugated oligo-nitroxide radicals (Scheme 1) have been employed as bridging ligands in which the spins of unpaired electrons interact ferromag-

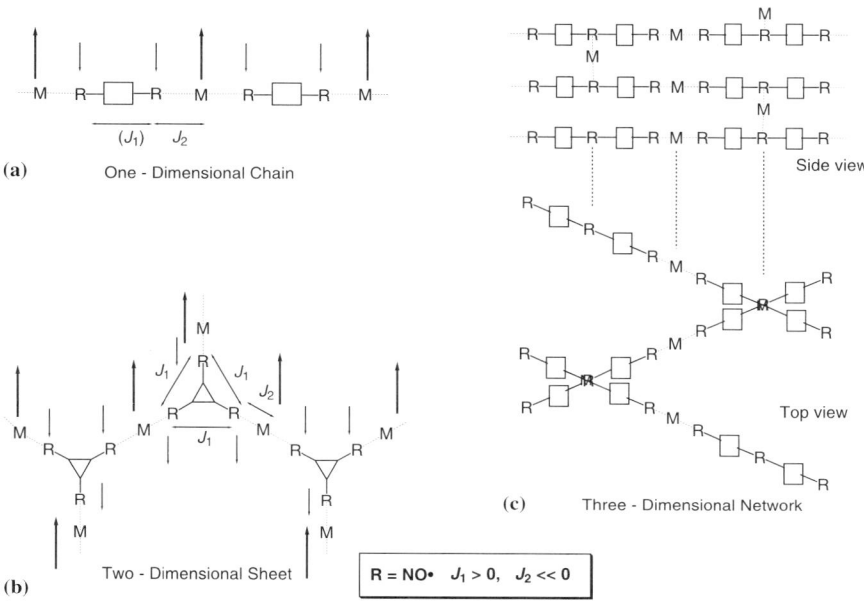

Fig. 5a–c. Schematic drawing of the coordination networks of the high-spin oligo-nitroxide radical and manganese systems

netically {*J*(intraligand) > 0}. Dimensions of the complexes, as well as sign and magnitude of the exchange coupling between the adjacent spins, may be readily tuned in this methodology [1, 3, 39–41].

The tris-monodentate triradicals **TNO**, **TNOPB**, **TNOPA**, **TNOB**, with quartet ground states ($S = 3/2$) in which the radical centers are arranged in a triangular disposition, would form 3:2 complexes with a coordinatively doubly unsaturated 3d metal ion M. In an ideal case a 2-D hexagonal network structure would be generated (Fig. 5b). A T-shaped quartet triradical carrying two non-equivalent ligating sites, e.g., **TNOP**, would form a 1-D chain by using two terminal nitroxide groups. The middle nitroxide group might then be used to cross-link the chains to form a 3-D network structure (Fig. 5c) depending on whether the second bridging takes place between the same chains as cross-linked by the first bridging. The spin alignment in these systems would be very much stabilized and is expected to give higher-T_C magnets.

2.3
Preparation of Metal-Poly(Nitroxide) Radical Complexes

The complex formation is a kinetically and/or thermodynamically controlled self assemblage of the reactants. Typical procedures are as follows. A suspension of manganese(II) bis(hexafluoroacetylacetonate) dihydrate, [Mn(hfac)$_2$ · 2H$_2$O], in *n*-heptane is heated at reflux to remove water of hydration by azeotropic distillation. To the resulting cooled solution is added **BNO$_H$** in *n*-heptane ether. The mixture is then concentrated under reduced pressure and the concentrated solution is allowed to stand to furnish black needles of [Mn(hfac)$_2$ · **BNO$_H$**] from a deep brown solution. It is preferable to carry out the reaction in an inert atmosphere and under anhydrous conditions, sometimes in a refrigerator. The reaction is completed by precipitation. Some complexes can be recrystallized but others dissociate in solution.

Excess of either component can generate complexes of different compositions. For example, the reaction of [Mn(hfac)$_2$] with tris(nitroxide) **TNOP** is complex while an equimolar mixture in ether containing *n*-hexane at −10 °C provides black blocks of a 1:1 complex [{Mn(hfac)$_2$} · **TNOP** · *n*-C$_6$H$_{14}$], a mixture containing a 1.7 molar excess of [Mn(hfac)$_2$] in *n*-heptane-ether generates black blocks of a 3:2 complex [{Mn(hfac)$_2$}$_3$ · **TNOP$_2$**] in ten days at 0 °C. The complex [{Mn(hfac)$_2$}$_3$ · **TNOPB$_2$** · *n*-C$_7$H$_{16}$] is obtained by dissolving [Mn(hfac)$_2$ · 2H$_2$O] in a mixture of diethyl ether, *n*-heptane, and benzene, followed by addition of **TNOPB** in benzene. Black blocks are formed from a deep violet solution. While [Mn(hfac)$_2$] gave similar black violet 3:2 complexes with tris(nitroxide) **TNOB**, but **TNO** and bis(nitroxide) **PNNNO** did not form polymeric complexes, probably due to steric hindrance around the ligand molecule.

Recently, a notorious side reaction has been elucidated leading to undesired by-products that have unique [3 + 3] benzene-dimer structures [42]. 1-D ferrimagnetic complexes, [Mn(hfac)$_2$ · **BNO$_X$**]*n* (X = Cl or Br), are typically obtained by the reaction of Mn(hfac)$_2$ with **BNO$_X$**. When crystallization requires a period of several days, the black solutions often turn yellow in about one day and do not afford the expected black polymer complexes. Instead, yellow crystalline precipitates are obtained under these conditions (Scheme 3).

An X-ray structure analysis revealed that **BBNO$_X$** · [Mn(hfac)$_2$ · H$_2$O]$_2$ · CH$_2$Cl$_2$ has a [3 + 3] benzene-dimer structure {X = Cl or Br; **BBNO$_X$** = 3, 10-dihalo-5, 8, 11, 12-tetrakis(*N-tert*-butylimino)-tricyclo[5, 3, 1, 1^{2,6}] dodeca-3, 9-diene *N,N′,N″,N‴*-tetraoxide} (Fig. 6). Two crystallographically equivalent manganese(II) ions have an octahedral coordination and are coordinated with four oxygen atoms of two hfac ligands, one oxygen atom of water, and one oxygen atom, O1, of **BBNO$_{Br}$**. Since the dimer complex is chiral, both enantiomers are contained within each unit cell (Scheme 4). The **BBNO$_{Cl}$** complex is isostructural to the bromine derivative.

The resonance structure **BNO$_{Br'}$** must be responsible for the reaction leading to the dimer complex (Scheme 5); either dimerization of **BNO$_{Br'}$** or attack of **BNO$_{Br'}$** to free and complexed **BNO$_{Br}$**. Whereas the 1-D ferrimagnetic complex [Mn(hfac)$_2$ · **BNO$_{Br'}$**]$_n$ is a kinetic product and precipitates from solution at an early stage of reaction, the yellow crystals of **BBNO$_{Br}$** · [Mn(hfac)$_2$ · H$_2$O]$_2$ · CH$_2$Cl$_2$ appear to be a thermodynamic product.

Another interesting feature of this work is the liberation of the dimer ligand **BBNO$_{Br}$**, free from manganese ions, by dissolving the complex in ether. Two water molecules of hydration appear to be crucial for the stability of **BBNO$_{Br}$** · 2H$_2$O; it is stable in water at 100 °C but starts to decompose by dissociation even at −78 °C in CH$_2$Cl$_2$ when dehydrated by molecular sieves.

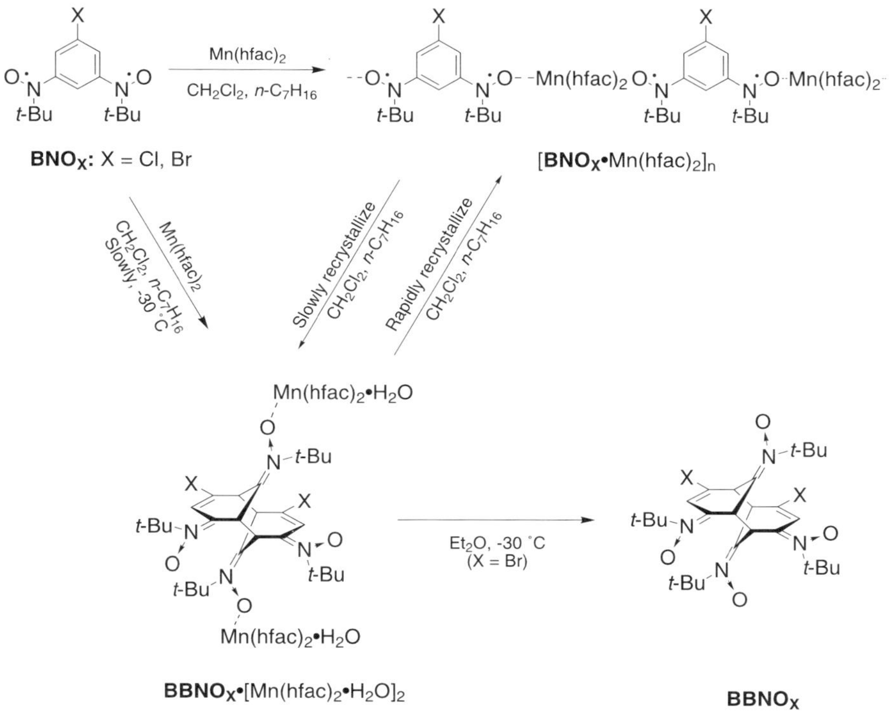

Scheme 3.

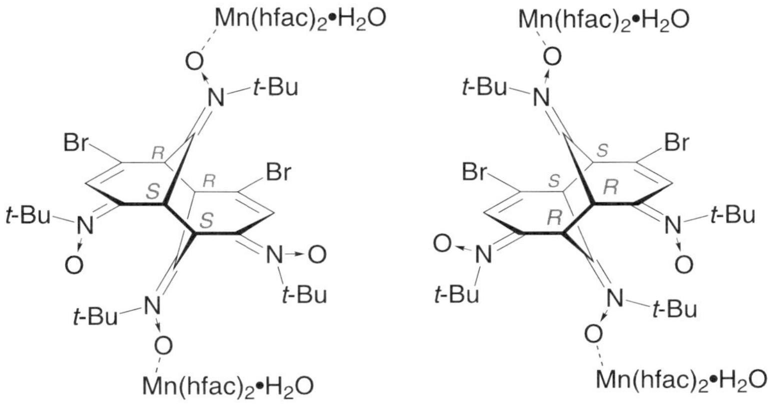

Scheme 4.

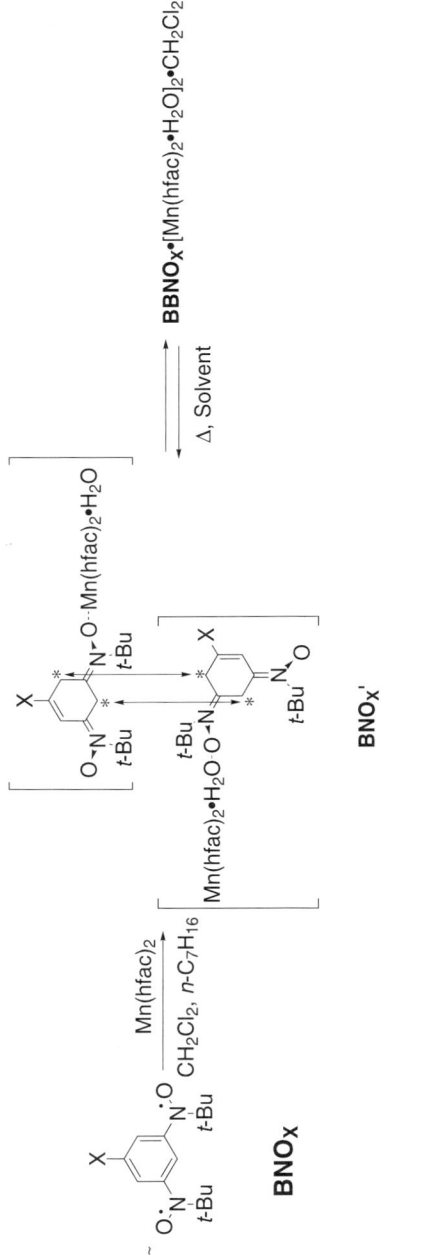

Scheme 5.

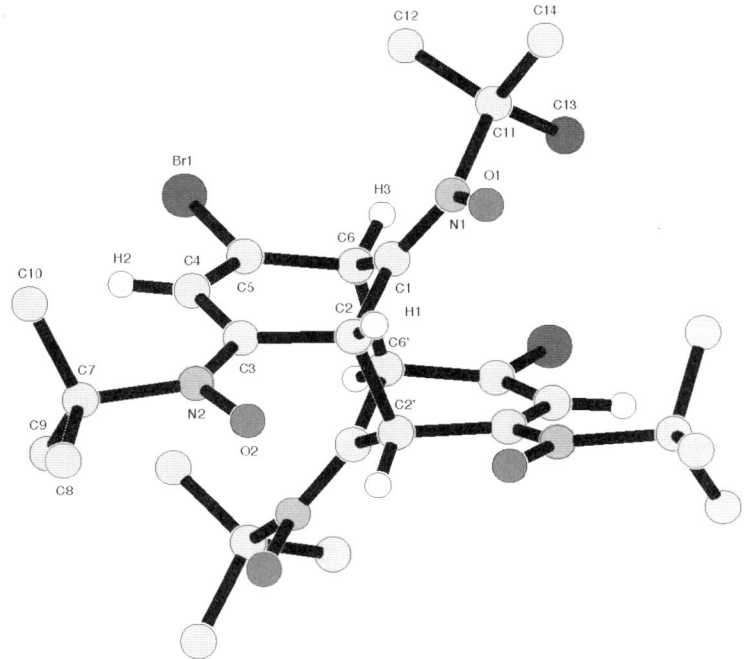

Fig. 6. Molecular structure of benzene-dimer **BBNO**$_{Br}$

3
One-Dimensional Nitroxide-Metal Systems

The first one-dimensional polymeric transition metal complex with nitroxides was [Cu(hfac)$_2$ · **TEMPOL**] [24]. Coupling between the nitroxide radical and copper(II) is ferromagnetic with $2J/k_B = 19 \pm 7$ K and this ($S = 1$) pair couples with adjacent pairs through quite lengthy superexchange paths of the σ-bonds of **TEMPOL** by $2J' = -78(2)$ mK. Since Gatteschi et al. reported a polymeric transition metal complex with nitronyl nitroxide (**NIT**) in 1986 [19, 43–56] others with 1,3-phenylenebis(nitroxide) derivatives (**BNO**$_R$) [1, 15, 30–33, 41, 57–61] have been documented (Table 1). Bis-monodentate nitronyl nitroxides with doublet states ($S = 1/2$) and bis-monodentate 1,3-phenylene-bis(nitroxides) with triplet ($S = 1$) ground states form 1:1 complexes having one-dimensional infinite chain structures with coordinatively doubly unsaturated paramagnetic metal ions. The coupling results in the alignment of neighboring spins in either parallel or antiparallel fashion to one another along the chain thereby giving one-dimensional ferromagnets or ferrimagnets, unless the sizes of the spins match each another in the latter.

The paramagnetic susceptibility values of both kinds of materials are expected to diverge at low temperatures, as a result of lengthening of the correlation of the spins along the chain. Depending on the nature of the

Table 1. Nitroxide-metal-based one-dimensional complexes

Formula	Type of sequence	$2J/k_B$ ($J_{interchain}/k_B$)	T_C or T_N (magnetism)	Ref.
Cu(hfac)$_2$(NITR) R = Me, i-Pr	—Cu—NIT—Cu—NIT—	$J_{CuNO}/k_B = 37.0$ K, $J_{interchain} < 0$ (R = Me), $J_{CuNO}/k_B = 35.4$ K, $J_{interchain} > 0$ (R = i-Pr)	$T_N < 1$ K, $T_C < 1$ K	[43, 46, 54, 19]
Mn(hafc)$_2$(NITR), R = Me, Et, i-Pr, n-Pr, Ph	—Mn—NIT—Mn—NIT—	$J_{MnNO}/k_B = -340$ to -475 K	$T_C = 7.6$ K (R = i-Pr), $T_C = 8.1$ K (R = Et), $T_C = 8.6$ K (R = n-Pr)	[45, 47], [50], [51]
Ni(hafc)$_2$(NITMe)	—Ni—NIT—Ni—NIT—	$J_{NiNO}/k_B = -610$ K		[48]
Eu(hafc)$_3$(NITEt)	—Eu—NIT—Eu—NIT—	$J_{EuNO}/k_B = -23.4$ K		[49]
Gd(hafc)$_3$(NITEt)	—Gd—NIT—Gd—NIT—		$T_C = 5.3$ K	[49, 54]
Zn(hafc)$_2$(NITi-Pr)	—Zn—NIT—Zn—NIT—	$J_{NONO}/k_B = -17.6$ K	(Antiferromagnetic chain)	[56]
Mn(hafc)$_2$(BNO$_H$)	—Mn—BNO$_H$—	$J_{MnNO}/k_B < -300$ K, $J_{TrTr}/k_B = +24$ K $(-0.11$ K$)$[a]	$T_N = 5.5$ K (Metamagnet)	[30, 62]
Mn(hafc)$_2$ (BNO$_F$)	—Mn—BNO$_F$—	$J_{MnNO}/k_B < -300$ K, $J_{TrTr}/k_B = +23$ K $(-0.06$ K$)$[a]	$T_N = 5.0$ K (Metamagnet)	[61]
Mn(hafc)$_2$(BNO$_{Cl}$)	—Mn—BNO$_{Cl}$—	$J_{MnNO}/k_B < -300$ K, $J_{TrTr}/k_B = +22$ K $(0.11$ K$)$[a]	$T_C = 4.8$ K	[60, 62]
Mn(hafc)$_2$(BNO$_{Br}$)	—Mn—BNO$_{Br}$—	$J_{MnNO}/k_B < -300$ K, $J_{TrTr}/k_B = +24$ K $(0.13$ K$)$[a]	$T_C = 5.3$ K	[60, 62]
Mn(hafc)$_2$(TNOP)	—Mn—TNOP—	—	$T_N = 11.0$ K (Metamagnet)	[33, 63]
Mn(hafc)$_2$(BNO$_{chiral}$)	—Mn—BNO$_{chiral}$—		$T_N = 5.4$ K (Metamagnet)	[15]
Mn(hafc)$_2$(NN$_{chiral}$)	—Mn—NN$_{chiral}$—	$J_{MnNO}/k_B = -435$ K ($J_{interchain}/k_B = 0.0069$ K)	$T_C = 4.5$ K (Ferrimagnet)	[16]

[a] Data were evaluated by quantum-classical approximation of ferromagnetic 3/2 spins. J_{MnNO} and J_{TrTr} define the interactions between Mn and NO and between two NO—Mn—NO's groups exhibiting $S = 3/2$ spins, respectively.

additional interchain interaction, the chain polymers become a bulk antiferromagnet or a ferri/ferromagnet. In these magnets the interchain magnetic interactions are much weaker than the intrachain interaction and, therefore, the transition temperatures to three-dimensionally ordered states are relatively low.

3.1
Structure and Magnetic Properties of Ferrimagnetic 1-D Chain Formed by Manganese(II) and Nitroxide Radicals

3.1.1
Manganese(II) Complexes with Nitronyl Nitroxides

Manganese(II) ions form crystalline chain complexes with various NIT_R (R = i-Pr, Et, n-Pr, P-alkyloxyphenyl) in which Mn(II) is hexacoordinated with four oxygen atoms of two hfac molecules and with two oxygen atoms of two different NIT radicals. The other oxygen atoms of the two NIT_R radicals are coordinated to the adjacent Mn(II) ions (Scheme 6). All the above complexes are ferrimagnetic and order at ca. 8 K. Quantitative analyses of the magnetic susceptibility data are performed utilizing a model of classical-quantum chains. The results are summarized in Table 1.

[Mn(hfac)$_2$] gave a crystalline 1-D chain complex with the NIT_{chiral} radical. The X-ray crystal structures of the NIT_{chiral} complex solved in the chiral space group $P2_12_12_1$ (No. 19). The asymmetric unit of [NIT_{chiral} · Mn(II)(hfac)$_2$]$_n$ consists of one Mn(II) ion, two hfac anions, and one chiral radical NIT_{chiral}. The Mn(II) ion and the NIT_{chiral} form a one-dimensional structure along the crystal a-axis. The saturation magnetization of this complex was 3.8 μ_B, which agrees well with the theoretical limit, assuming antiferromagnetic coupling between the manganese and radical spin. This complex is ferrimagnetic and orders at 4.5 K [16].

3.1.2
Manganese(II) Complexes with Triplet Bis-Nitroxide Radicals

Manganese(II) ions couple form crystalline chain complexes with triplet bis-nitroxide benzene derivatives [30–33]. X-ray crystal structures of the achiral BNO_R complexes solved in the monoclinic $P2_1/n$ space group (No. 14) with Z = 4 reveal that the manganese(II) ions have an octahedral coordination with four oxygen atoms of two hfac anions and two oxygen atoms of two different BNO_R molecules. The latter is bound to the Mn(II) ion in a *cis*-configuration.

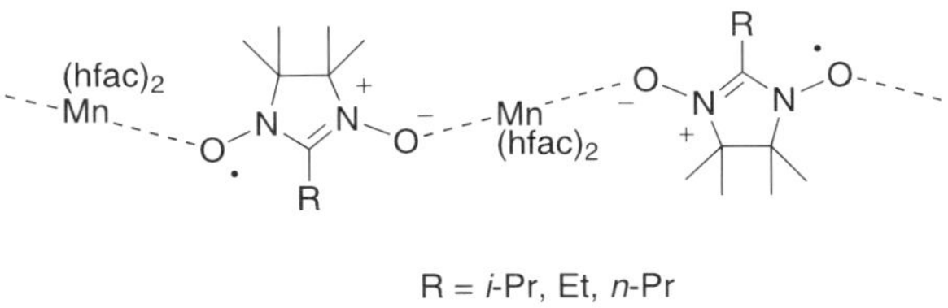

R = i-Pr, Et, n-Pr

Scheme 6.

As a result, the Mn ions and biradical molecules form a helical 1-D polymeric chain structure along the crystal b axis. All the hexacoordinated Mn(II) ions have either the Δ- or the Λ-configuration along a given chain. Two *tert*-butylnitroxide groups are rotated out of the phenylene ring plane in a conrotatory manner, but with different angles; each BNO_R molecule in the crystal has no symmetry element and is therefore chiral, i.e., R or S. The 1-D polymeric chains are therefore isotactic as all units of the same chirality reside on a given chain (Fig. 7a). The crystal lattice is, as a whole, achiral since an enantiomeric chain is present. In the case of the chiral BNO_{chiral} ligand, the crystal lattice is, as a whole, chiral (Fig. 7b) [15].

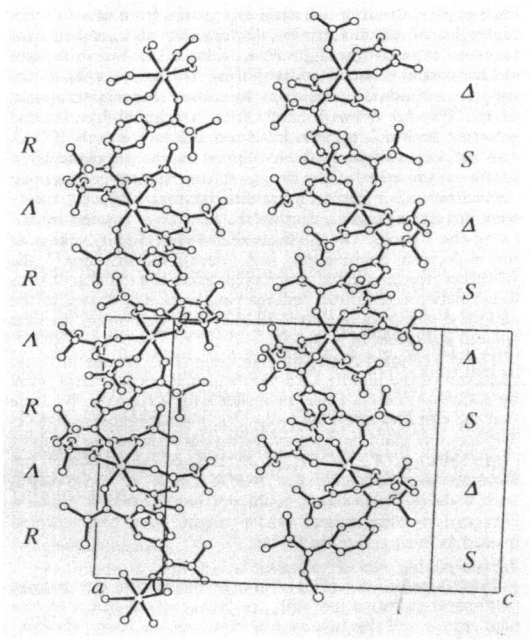

The observed $\chi_{mol}T$ value at 300 K varies between 2.2 and 2.4 emu K mol^{-1}, which is slightly larger than the theoretical limit of 1.88 emu K mol^{-1} expected for $S = 3/2$ ferromagnetic chain compounds. As T is lowered, the $\chi_{mol}T$ value increases monotonically in proportion to the increase in the correlation length within the network. Together with the lack of a minimum at lower temperature, the room temperature $\chi_{mol}T$ value also points to the operation of strong (more negative than -300 K) antiferromagnetic coupling between the Mn(II) ion and the nitroxide radical of **BNO$_R$**, in which the onset of the intramolecular ferromagnetic coupling is meaningful.

The saturation magnetization values, $M_S = 3\ \mu_B$/f.u., of all 1-D of Mn(hfac)$_2$(**BNO$_R$**) complexes agree well with the theoretical limit, assuming antiferromagnetic coupling between the manganese and triplet radical spins.

The low-field susceptibilities of the Mn(hfac)$_2$(**BNO$_H$**) and Mn(hfac)$_2$(**BNO$_F$**) complexes show a sharp cusp at 5.5 K and 4.5 K respectively (Fig. 8). This behavior indicates that these complexes possess an antiferromagnetic ground state. On the other hand, the Mn(hfac)$_2$(**BNO$_{Cl}$**) and Mn(hfac)$_2$(**BNO$_{Br}$**) show demonstrate ferrimagnetic behavior (Fig. 8).

Magnetization at 1.8 K revealed metamagnetic behavior for the Mn(hfac)$_2$(**BNO$_H$**) and Mn(hfac)$_2$(**BNO$_F$**) complexes (Fig. 9). Namely, although the response of the magnetization was not sensitive to the weak applied magnetic field below H_C, a behavior characteristic of an antiferromagnet, a sharp rise and approach to saturation of magnetization, characteristic of a ferromagnet, was observed at higher applied magnetic field. The complexes Mn(hfac)$_2$(**BNO$_{Cl}$**) and Mn(hfac)$_2$(**BNO$_{Br}$**) show ferrimagnetic behavior (Fig. 9). These compounds show narrow hysteresis with coercive force being less than 20 Oe (Inset of Fig. 9).

3.2
Conclusion

Mn complexes with **BNO$_{Cl}$** and **BNO$_{Br}$** have the same crystal structure as the complexes with **BNO$_H$** and **BNO$_F$**. From the observed intermolecular distance, the strongest interchain interaction in Mn(hfac)$_2$(**BNO$_R$**) appears to arise from the Mn—Mn and Mn—C$_{middle}$ distances. From the dipole-dipole interaction term, the magnetic interaction between nearest neighbor magnetic moments is always antiferromagnetic. This argument is confirmed by the fact that $2J'/k_B$ does not exhibit a regular change throughout the Mn(hfac)$_2$(**BNO$_R$**) series, but changes abruptly in sign when the nearest interacting pair is altered. Consequently, the ferri/ferromagnet ground state is stabilized in complexes containing **BNO$_{Cl}$** and **BNO$_{Br}$** (Fig. 10a), whereas complexes containing **BNO$_H$** and **BNO$_F$** behave as antiferromagnets (Fig. 10b). Due to a weak interchain exchange interaction, the antiferromagnetic complexes Mn(hfac)$_2$(**BNO$_H$**) and Mn(hfac)$_2$(**BNO$_F$**) display metamagnetic behavior characteristic of highly anisotropic magnetic materials.

◆

Fig. 7. (a) View of a 1-D chain formed by a bisnitroxide **BNO$_H$** radical and Mn(II)(hfac)$_2$. (b) View of a chiral 1-D chain formed by a bisnitroxide **BNO$_{chiral}$** radical and Mn(II)(hfac)$_2$

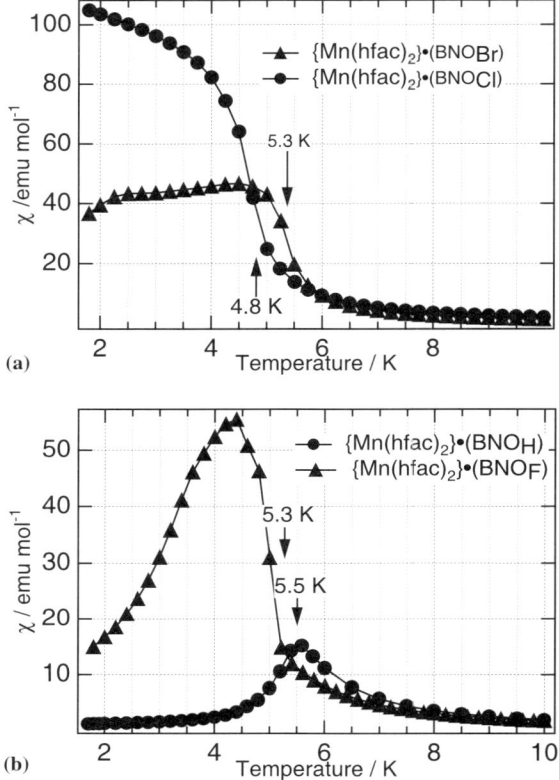

Fig. 8. Temperature dependence of the low-field susceptibility (5 Oe) of the ferrimagnetic complexes (a) Mn(hfac)$_2$ · (**BNO$_H$**) and Mn(hfac)$_2$ · (**BNO$_F$**) and (b) Mn(hfac)$_2$ · (**BNO$_{Cl}$**) and Mn(hfac)$_2$ · (**BNO$_{Br}$**)

The magnetic properties of chiral 1-D polycrystals of [**NIT$_{chiral}$** · Mn(II)(hfac)$_2$]$_n$ are similar to those of the ferrimagnetic complexes [**BNO$_{Cl}$** · Mn(II)(hfac)$_2$]$_n$ and [**BNO$_{Br}$** · Mn(II)(hfac)$_2$]$_n$. The magnetic properties of polycrystals of chiral 1-D magnets of [**BNO$_{chiral}$** · Mn(II)(hfac)$_2$]$_n$ are similar to those of the metamagnetic complexes [**BNO$_H$** · Mn(II)(hfac)$_2$]$_n$ and [**BNO$_F$** · Mn(II)(hfac)$_2$]$_n$.

4
Two-Dimensional Metal-Nitroxide Systems

In principle, there are no difficulties in the design of materials that undergo magnetic phase transitions at higher temperatures: the magnetic network of coupled metal ions and radicals should be extended from one to two or three dimensions, and the strong magnetic coupling between the spin centers would

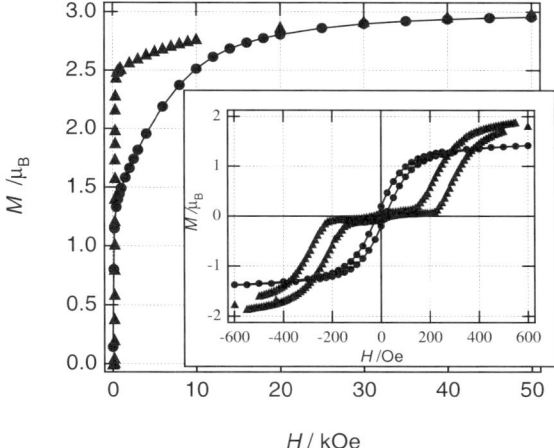

Fig. 9. Magnetization curves of Mn(hfac)$_2$ · (**BNO$_{Br}$**) (●) and Mn(hfac)$_2$ · (**BNO$_H$**) (▲) at 1.7 K. Inset shows the low-field cycling

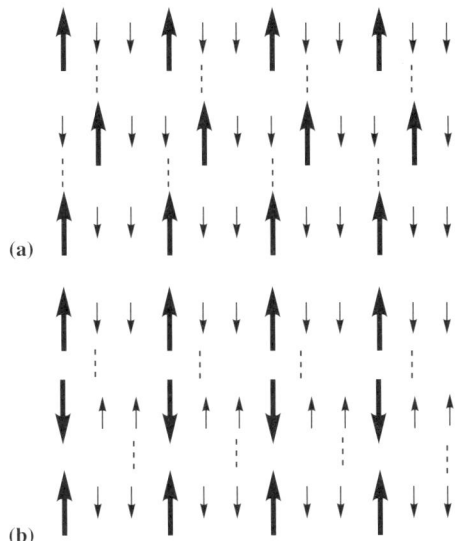

Fig. 10. Schematic drawing of the magnetic structure of (**a**) Mn(hfac)$_2$ · (**BNO$_{Cl}$**) and Mn(hfac)$_2$ · (**BNO$_{Br}$**) and (**b**) Mn(hfac)$_2$ · (**BNO$_H$**) Mn(hfac)$_2$ · (**BNO$_F$**). Broken lines represent the antiferromagnetic interaction between the ferro/ferrimagnetic 1-D chains. Solid lines represent the ferromagnetic interchain interaction

ensure high transition temperatures [64]. One of the first two-dimensional structures is a series of copper(II) β-diketonate complexes [65–71]. The coordination of the oxygen atoms of the nitroxide groups of neighboring molecules completes the distorted octahedra around the Cu(II) ions. Temperature dependence of the magnetic susceptibility of these complexes fits a theory of isolated ferromagnetic coupled pairs. Enaminoketones A and B of 3-imidazoline-1-oxyl form layered polymeric structures [72–75] with Cu(II), Co, and Ni. However, while all these complexes exhibit extended structures, the magnetic coupling is limited to the directly-bonded metal-nitroxide couples.

Thus, the problem is not only how to control the construction and the structure of extended systems in a desired fashion in order to extend the properties from the individual building blocks to the entire lattice, but also how to maintain strong magnetic coupling throughout the extended structure. Moreover, there is much difficulty in making a single crystal of the complex which has high dimensionality. Therefore, there are still only few examples of nitroxide-metal ion-based two-dimensional magnetic complexes (Table 2) [14, 31, 32, 39, 52, 76, 77]. The flexibility of the ligands is important to make good crystals of two- or three-dimensional complexes. In metal-nitronyl nitroxide systems, there are limitations with regard to the topological synthesis of two- or three-dimensional complexes. The approach through high spin oligo-nitroxides and transition metal complexes is more advantageous than that through metal-nitronyl nitroxide systems.

4.1
Structure and Magnetic Properties of Ferrimagnetic 2-D Sheet Formed by Manganese(II) and Nitroxide Radicals

4.1.1
2-D Manganese(II) Complexes with Nitronyl Nitroxides

Some ferrimagnetic 2-D manganese(II)-nitronyl nitroxide systems were synthesized by the complexation of manganese(II) bispentafluorobenzoate

Table 2. Nitroxide-metal-based two-dimensional complexes

Formula	Crystal or powder	J/k_B ($J_{interlayer}/k_B$)	T_C or T_N	Ref.
{Mn(pfbz)$_2$}$_2$ (NITR), R = Me,Et	Powder		T_C = 24 K, T_C = 20.5 K	[52]
{Mn(pfpr)$_2$}$_2$ (NITR), R = Me,Et	Powder	−149.6 K (0.075 K)	T_C = 24 K, T_C = 20.5 K	[76]
{Mn(hafc)$_2$}$_3$ (NITBzald)$_2$ · 0.5CHCl$_3$	Powder	−367 K	T_C = 6.4 K	[77]
{Mn(hafc)$_2$}$_3$ (TNOPB)$_2$ · n-C$_7$H$_{16}$	Crystal	J_{MnNo}/k_B = −175.4 K, J_{NONO}/k_B = 0.225 K[a] (0.002 ± 0.001 K[a])	T_C = 3.4 K	[31, 32, 39, 78]
{Mn(hafc)$_2$}$_3$(TNOP)$_2$	Crystal	J_{MnNo}/k_B < −350 K	T_C = 9.5 K	[14, 79]

[a] Data were evaluated by quantum-classical approximation of ferromagnetic 3/2 spins. J_{MnNo} and J_{NONO} define the interactions between Mn and NO and between two NO spins, respectively.

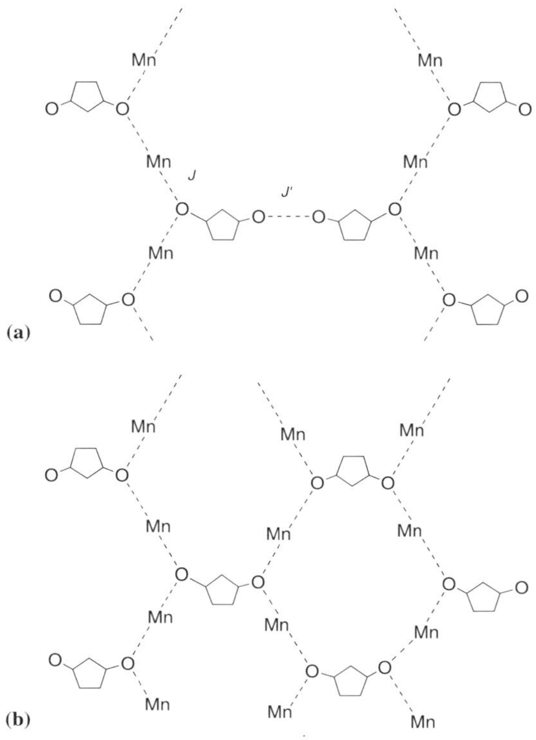

Scheme 7.

[Mn(pfbz)$_2$, pfbz = pentafluorobenzoate] or manganese(II) bispentafluoropropionate [Mn(pfpr)$_2$, pfpr = pentafluoropropionate] with 2-alkylnitronyl nitroxide (**NIT$_R$**, R = Me, Et) (Scheme 7). None of the complexes gave single crystals amenable for an X-ray crystal structure analysis. It is assumed that 2-D sheet structures were formed in these complexes. The complexes of [Mn(pfbz)$_2$]$_2$(**NIT$_R$**) (R = Me, Et) and [Mn(pfpr)$_2$]$_2$(**NIT$_R$**) (R = Me, Et) are ferrimagnetic and the transition into an ordered state occurs at ca. 23 K. The quantitative analysis of the magnetic susceptibility of these complexes was performed by using a model of a Heisenberg classical-quantum spin system (see Table 2).

4.1.2
Manganese(II) Complexes with Triplet Bis-Nitroxide Radicals

The X-ray crystal structure of the complex {Mn(hfac)$_2$}$_3$ · (**TNOPB**)$_2$ · n-C$_7$H$_{16}$ revealed an isostructure and that the manganese(II) ion has an octahedral coordination with the four oxygen atoms of two *hfac* anions bound to the metal ion in the equatorial plane while the axial positions are occupied

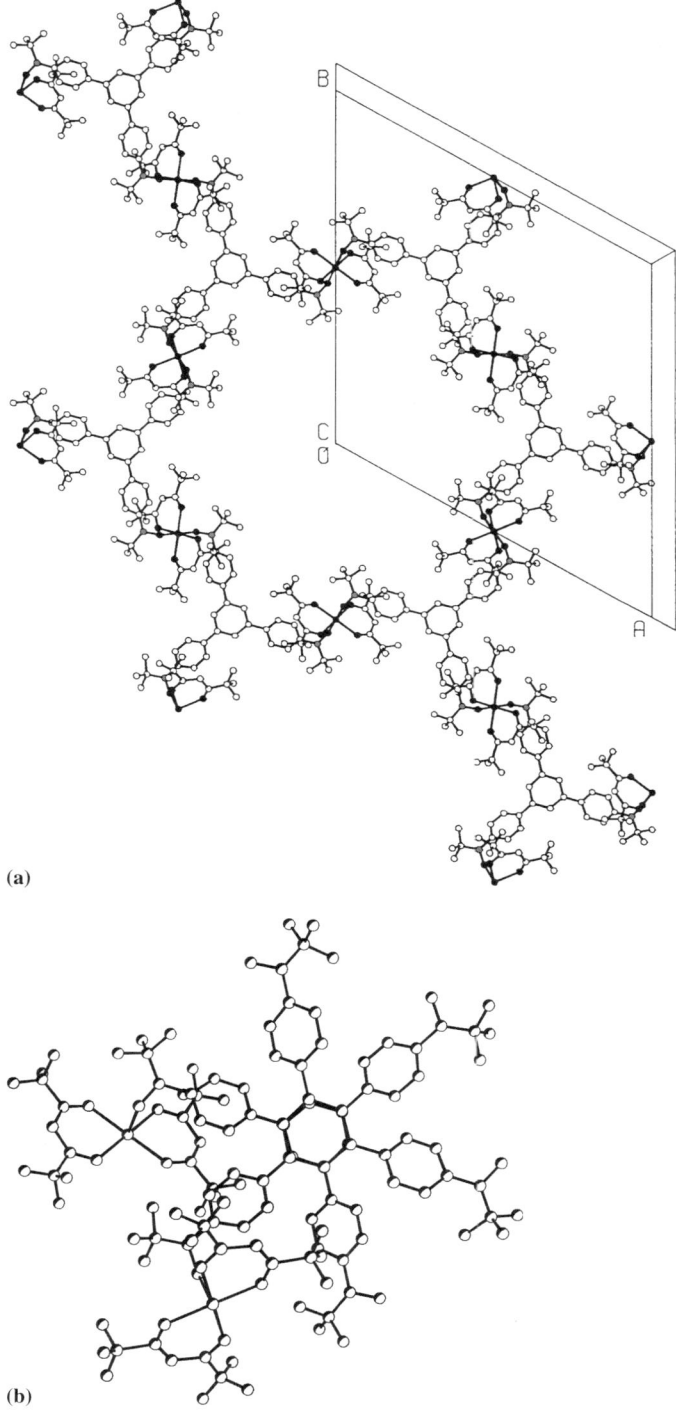

(a)

(b)

Fig. 11. (a) View along the c axis of a layer displaying hexagons comprised of six trinitroxide **TNOPB** and six manganese complexes Mn(hfac)2. (b) View of the shortest contact between the layers

by the two oxygen atoms of the two nitroxide groups. Six triradical molecules and six Mn ions form a hexagon from which an extended honeycomb network is constructed by sharing edges (Fig. 11a). A disordered n-heptane molecule is contained within each hexagonal cavity. The 2-D network sheets form a layered structure in which the adjacent layers can slide by a radius of the hexagon from the superimposable disposition with a mean inter-plane distance was 3.58 Å (Fig. 11b). On the basis of spin density and distance, the strongest inter-plane spin-spin interaction is considered to arise from the carbons (3.78 Å apart) of the benzene rings *para* and *meta* to the nitroxide groups (Fig. 11b). This type of interaction is expected to be ferromagnetic as dictated by McConnell's theory [80, 81].

While satisfactory analytical data were obtained for the complex {Mn(hfac)$_2$}$_3$ · (**TNOB**)$_2$ [14], the complex did not afford a single crystal amenable to X-ray crystal structure analysis. It is assumed that, while a 2-D sheet structure is formed, the reduced symmetry of **TNOB** relative to **TNOPB** must be responsible for the difficulty in obtaining single crystals. Magnetic and analytical data support the 2-D nature of this complex. The 2-D network is schematically presented in Fig. 12. As in the case of **TNOB**, the network may be stacked ferromagnetically across layers.

The magnetic susceptibility of {Mn(hfac)$_2$}$_3$(**TNOPB**)$_2$ · n-C$_7$H$_{16}$ and {Mn(hfac)$_2$}$_3$(**TNOB**)$_2$ were investigated in the temperature range 1.8–350 K and fields up to 5 Tesla. For the complex {Mn(hfac)$_2$}$_3$(**TNOPB**)$_2$ · n-C$_7$H$_{16}$ with increasing temperature, this dependence is first characterized by a sharp maximum at 2.5 K followed by a smooth minimum in the range 80–140 K, which rather resembles a plateau. The minimum value of $\chi_{mol}T = 5.71$ emu K/mol within this interval is very close to the theoretical limit of 5.625 emu K/mol which is expected for three stable non-interacting $S = 3/2$ spins per mole. According to Fig. 11a, this spin configuration can be considered as being formed by one $S = 5/2$ spin of the Mn(II) ion coupled antiferromagnetically with two 1/2-spins of two different **TNOPB** triradicals.

For the complex {Mn(hfac)$_2$}$_3$(**TNOB**)$_2$, the $\chi_{mol}T$ values increased with decreasing temperature, displaying a maximum at 10 K. The value of $\chi_{mol}T = 7.1$ emu K/mol is larger than the theoretical limit of 5.625 emu K/mol, which is expected for three stable non-interacting $S = 3/2$ spins per mole. According to Fig. 12, this spin configuration can be considered as being formed by one $S = 5/2$ spin of the Mn(II) ion coupled antiferromagnetically with two 1/2-spins of two different **TNOB** triradicals. Together with the lack of a minimum at lower temperature, the room temperature $\chi_{mol}T$ value also points to the operation of strong (more negative than -300 K) antiferromagnetic coupling between the Mn(II) ion and the nitroxide radical

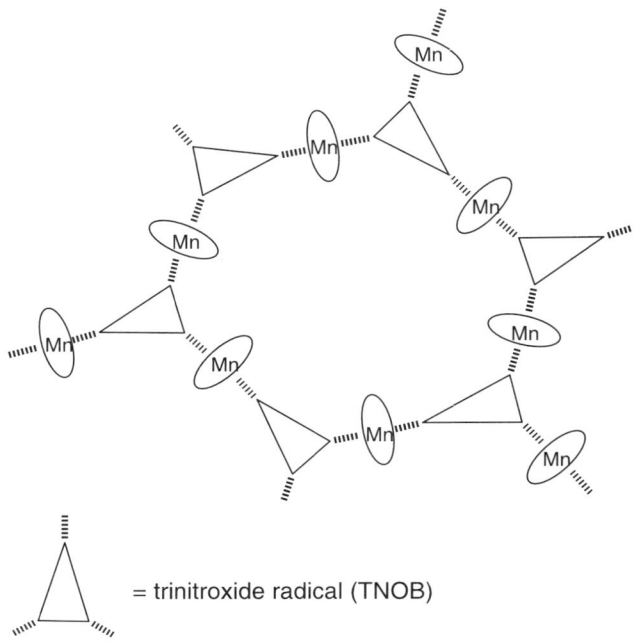

Fig. 12. A schematic drawing of the estimated 2-D network structure of {Mn(hfac)$_2$}$_3$ · (**TNOB**)$_2$

of **TNOB**, in which the onset of intramolecular ferromagnetic coupling is meaningful.

When measurements were conducted in a low field, the magnetization values of {**Mn(hfac)**$_2$}$_3$ · (**TNOPB**)$_2$ · n-C$_7$H$_{16}$ and {**Mn(hfac)**$_2$}$_3$ · (**TNOB**)$_2$ complexes show a sharp rise at T_C = 3.4 K and 9.2 K, respectively (Fig. 13). The low-field susceptibilities below T_C are extremely large, as expected for a ferro/ferrimagnet. Spontaneous magnetization was observed below T_C, demonstrating the transition to a bulk magnet. Saturation magnetization values of these two 2-D complexes (M_S = 9 μ_B/mol.) agree closely with the theoretical limit assuming antiferromagnetic coupling between the manganese and triplet radical spins (Fig. 14). These compounds exhibit narrow hysteresis with coercive force of less than 20 Oe.

4.2
Conclusion

It has been shown that perfect 2-D ferrimagnetic sheets exhibiting heterospin ferromagnetic ($J_1 > 0$)-antiferromagnetic ($J_2 < 0$ in Fig. 5b) networks together with ferromagnetic stacking of layers are realized in metal-nitroxide radical systems.

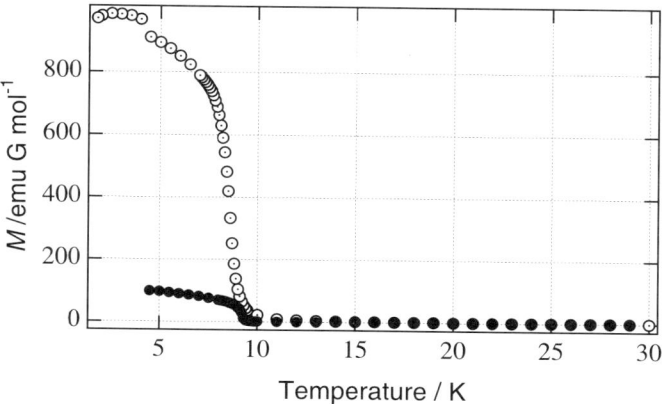

Fig. 13. Observed magnetization versus T plots for the complex $\{Mn(hfac)_2\}_3 \cdot (TNOB)_2$. Measurements were obtained at a magnetic field of 1 Oe (○) and spontaneous magnetization (●)

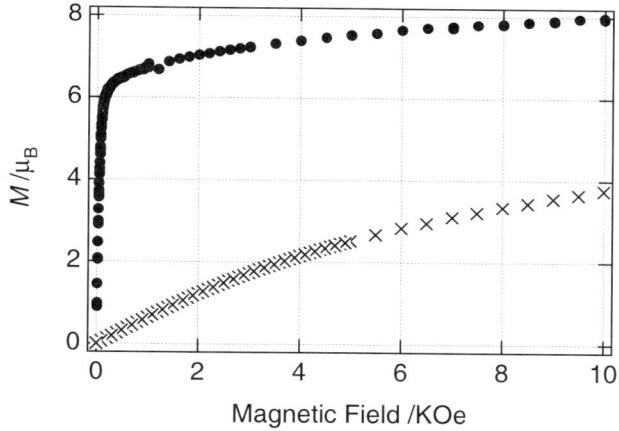

Fig. 14. Field dependence of the magnetization of $\{Mn(hfac)_2\}_3 \cdot (TNOB)_2$ measured at 1.8 K (○) and 10.8 K (+)

While the kinetic instability of the triradical **TNOB** did not allow us to study the magnitude of the intramolecular ferromagnetic exchange coupling, it is expected to be considerably greater for **TNOPB**. In cases involving **TNOB**, the interaction forms by its symmetry with respect to an isosceles triangular geometry. The value for a **TNOB** moiety located between the two nitroxide radicals at positions 3 and 5 of the same benzene ring is estimated from that of m-phenylenebis(N-$tert$-butylnitroxide) and analogues to be $J/k_B = 200$–500 K. The other interaction between the 4′ and 3 (and 5) nitroxide radical centers is estimated to be ca. 67 K. Altogether, the intramolecular ferromagnetic coupling in **TNOB** is estimated to be stronger in **TNOPB** and contributes to the higher T_C value of 9.2 K, in comparison to 3.4 K in the latter, where self-assembly with the assistance of Mn(II) ions is observed.

5
Three-Dimensional Metal-Nitroxide Systems

In Sects. 3 and 4 we described the one- and two-dimensional complexes, respectively. For one-dimensional complexes the transition temperatures to ferri/ferromagnets are about 5–6 K, while for two-dimensional complexes, T_C rises to ca. 9 K. In order to synthesize complexes with higher transition temperatures, the dimensionality of the spin network must be raised. The flexibility of ligands is more important in the construction of suitable crystals in three-dimensional complexes than in two-dimensional systems. Owing to these difficulties, in metal-nitroxide systems only three-dimensional complexes have been reported.

5.1
Crystal and Molecular Structures

An X-ray crystal structure analysis of an orthorhombic crystal of the complex revealed the formation of a parallel cross-shaped 3-D polymeric network (Fig. 15). The oxygen atoms of the terminal nitroxide groups of the triradical **TNOP** are ligated to two different manganese ions to form a 1-D chain in the *b/c* plane of the crystal. Since any manganese ion in an octahedral position is attached to the two nitroxide oxygens from two different triradical molecules in a *trans* disposition, the trinitroxide molecules are in a zigzag orientation along the chain. The diphenyl nitroxide unit is in a chiral propeller conformation and the R- and S-forms alternate along the chain. The middle nitroxide group of the ligand molecule **TNOP** on one chain is used to link its oxygen with that of the same chirality occurring in adjacent chains extended in the *b/-c* diagonal direction through a third Mn(II) ion in an octahedral position with an intersecting angle of 54.4°, thus establishing a parallel cross-shaped 3-D polymeric network (Fig. 16).

5.2
Magnetic Properties of 3-D Systems

The $\chi_{mol}T$ value of 8.64 emu K mol^{-1} at 300 K is larger than the theoretical value of 5.63 emu K mol^{-1} expected for short-range antiferromagnetic ordering of six 1/2 spins of **TNOP** and three 5/2 spins of d^5 Mn(II) for {Mn(hfac)$_2$}$_3$(**TNOP**)$_2$. As T is lowered, the $\chi_{mol}T$ value increases monotonically in proportion to the increase in the correlation length within the network. Together with the lack of a minimum at lower temperature, the room

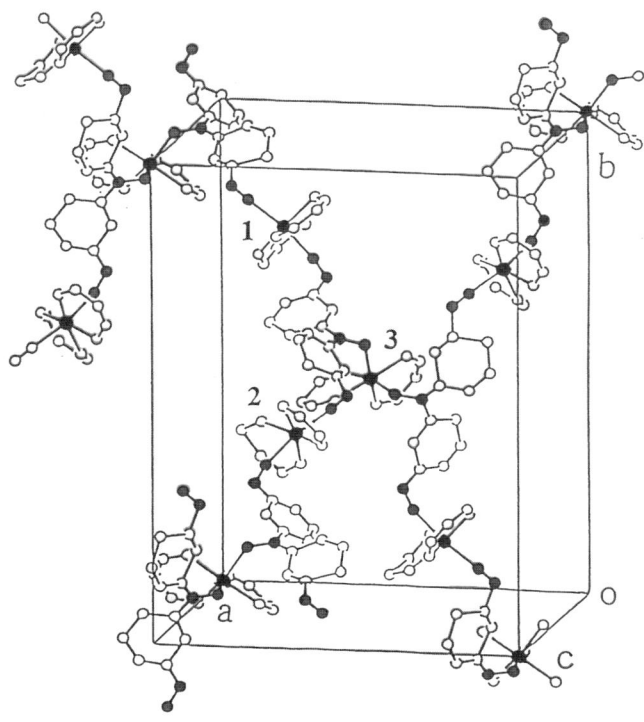

Fig. 15. Crystal structure of the 3-D metal-radical complex {Mn(hfac)$_2$}$_3$ · (**TNOP**)$_2$. The CF$_3$ and (CH$_3$)C groups are omitted for clarity. a, b and c denote the orthorhombic crystal axes. The Mn(*1*) and Mn(*2*) ions are represented by closed circles

temperature $\chi_{mol}T$ value points to the operation of strong (more negative than −300 K) antiferromagnetic coupling between the Mn(II) ion and the nitroxide radical of **TNOP**, in which the onset of the intramolecular ferromagnetic coupling is meaningful.

The temperature dependence of the magnetization M for a single crystal sample of [{Mn(hfac)$_2$}$_3$(**TNOP**)$_2$] was investigated at 5 Oe. When the sample was cooled within the field of 5 Oe, field-cooled magnetization displayed an abrupt rise at $T_C = 46$ K (Fig. 17).

The field dependence of magnetization at 5 K showed two important features. First, the magnetization rose sharply at low field reached a value of ca. 9 μ_B (50,000 emu G mol^{-1}) at 220 Oe, and became saturated. The saturation value is in good agreement with the theoretical value of 9 μ_B (5/2 × 3 − 3/2 × 2 = 9/2) expected for the antiferromagnetic coupling between the d^5 Mn(II) ion and the $S = 3/2$ triradical TNOP. Secondly, a conspicuous magnetocrystalline anisotropy was found in which the easy axis of magnetization lies along the c axis of the crystal lattice and the hard axis lies perpendicular to it (Fig. 18) [62, 63].

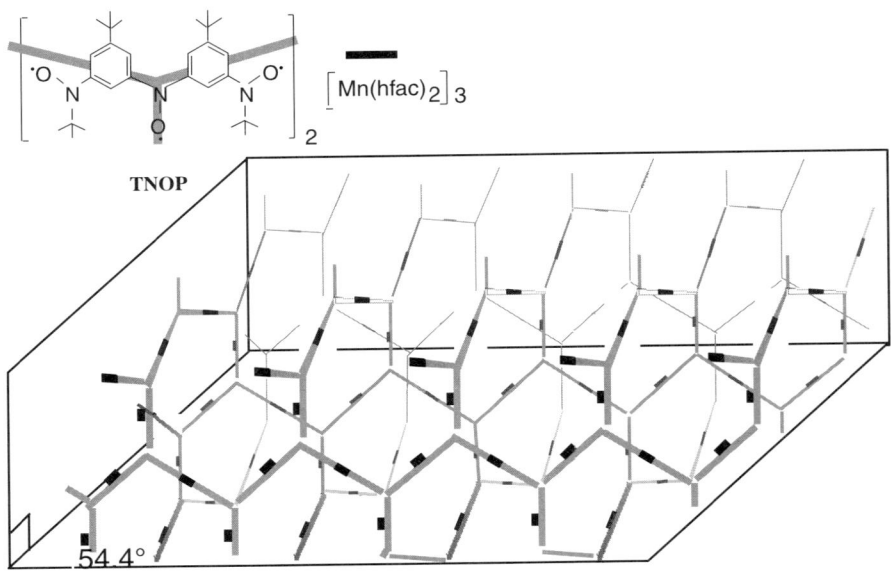

Fig. 16. Schematic drawing of a crystal of the 3-D metal-radical complex {Mn(hfac)$_2$}$_3$ · (**TNOP**)$_2$

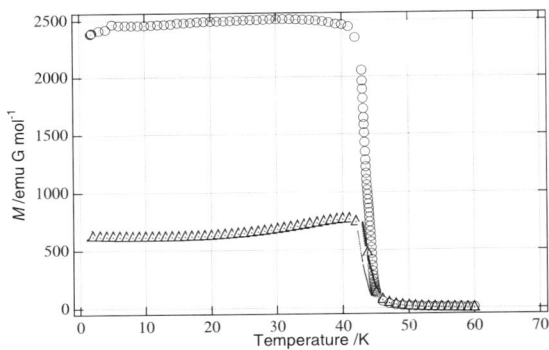

Fig. 17. Temperature dependence of magnetization of a single crystal of {Mn(hfac)$_2$}$_3$ · (**TNOP**)$_2$. (**a**) Zero field cooled magnetization (ZFC, ○) along the c-axis. (**b**) Zero field cooled magnetization (ZFC, △) along the b-axis

5.3
Conclusion

It is concluded that the perfect 3-D ferro/ferrimagnet with a heterospin ferromagnetic ($J_1 > 0$)-antiferromagnetic ($J_2 < 0$) network is realized in {Mn(hfac)$_2$}$_3$(**TNOP**$_2$). The saturation value is indicative for the antiferromagnetic coupling between the d^5 Mn(II) ion and the $S = 3/2$ triradical **TNOP**.

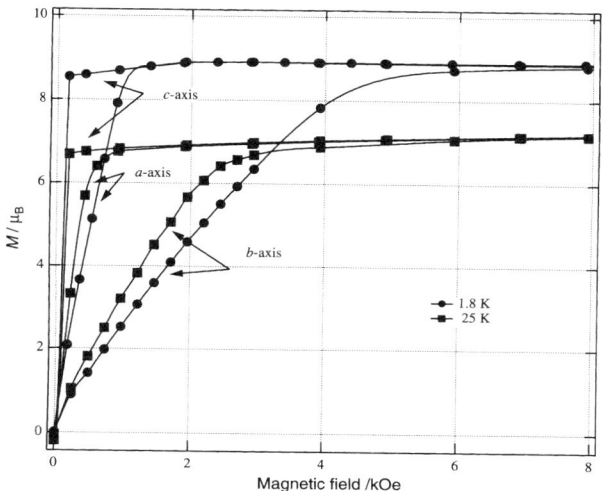

Fig. 18. Magnetization curves of {Mn(hfac)$_2$}$_3$ · (**TNOP**)$_2$ at 1.8 and 25 K along the three principal crystallographic axes

6
Conclusions

Supramolecular approaches have been successfully applied to the construction of high-dimensional network structures which are otherwise difficult to construct. These techniques utilize intermolecular interaction considerably weaker than the conventional covalent bond, e.g., van der Waals forces, hydrogen bonding, hydrophobic interaction, etc. The products are formed under thermodynamic, rather than kinetic control. Such network structures are intended for use as molecular switches, host molecular cages like "organic zeolites", and enzymatic pockets, electric conductors, non-linear optical materials, etc. As far as the magnetic properties are concerned, the exchange couplings through these weaker bonds are rather weak. For the coupling to be strong, unpaired electrons have to be bound to each another through one or two σ-bonds or several π-bonds. Thus, the supramolecular approaches have been limited to metal coordination compounds. Design of appropriate radicals is key to our strategy of employing high-spin oligo-nitroxide radicals as bridging ligands.

7
References

1. Iwamura H, Inoue K, Hayamizu T (1996) Pure Appl Chem 68: 243–252
2. Iwamura H (1990) Adv Phys Org Chem 26: 179–253
3. Iwamura H (1998) J Phys Org Chem 11: 299–304

4. Lim Y, Drago RS (1972) Inorg Chem 11: 1334–1338
5. Boymel PM, Eaton GR, Eaton SS (1980) Inorg Chem 19: 727–735
6. Boymel PM, Barden GA, Eaton GR, Eaton SS (1980) Inorg Chem 19: 735–739
7. Bilgren PM, Barden GA, Eaton GR, Eaton SS (1985) Inorg Chem 24: 4268–4272
8. Drago RS, Kuechler TC, Kroeger M (1979) Inorg Chem 18: 2337–2342
9. Dickman MH, Doedens RJ (1981) Inorg Chem 20: 2677–2681
10. Calder A, Forrester AR, James PG, Luckhurst GR (1969) J Am Chem Soc 91: 3724–3727
11. Ishida T, Iwamura H (1981) J Am Chem Soc 113: 4238–4241
12. Kanno F, Inoue K, Koga N, Iwamura H (1993) J Phys Chem 97: 13267–13272
13. Inoue K, Iwamura H (1995) Angew Chem Int Ed Engl 34: 927–928
14. Inoue K, Iwamura H (1996) Adv Mater 8: 73–76
15. Kumagai H, Inoue K (1999) Angew Chem Int Ed 38: 1601
16. Kumagai H, Inoue K (to be published)
17. Hosokoshi Y, Nakazawa Y, Inoue K, Takizawa K, Nakano H, Takahashi M, Goto T (1999) Phys Rev B 60: 12924
18. Hayami S, Inoue K (1999) Chem Lett 545–546
19. Caneschi A, Gatteschi D, Sessoli R, Rey P (1989) Acc Chem Res 22: 392–398
20. Caneschi A, Gatteschi D, Laugier J, Pardi L, Rey P (1988) Inorg Chem 27: 1031
21. Gatteschi D, Rey P (1999) In: Lahti PM (ed), Magnetic Properties of Organic Materials. Marcel Dekker, New York
22. Mushin RN, Schastnev PV, Malinovskaya SA (1992) Top Mol Organ Eng 9: 167–173
23. Mushin RN, Schastnev PV, Malinovskaya SA (1992) Inorg Chem 31: 4118
24. Cazianis CT, Eaton ER (1974) Can J Chem 52: 2454–2462
25. Dickman MH, Porter LC, Doedens RJ (1986) Inorg Chem 25: 2595–2599
26. Caneschi A, Gatteschi D, Sessoli R, Hoffman SK, Laugier J, Rey P, Sessoli R, Zanchini C (1988) J Am Chem Soc 110: 2795–2799
27. Pervukhina NV, Podlberezskaya NV, Ovcharenko VI, Larionov SV (1991) Zh Strukt Khim 32: 123–130
28. Rabu P, Drillon M, Iwamura H, Goerlitz G, Itoh, Matsuda K, Koga N, Inoue K (2000) Eur J Inorg Chem 211–216
29. Goerlitz G, Hayamizu, Itoh T, Matuda K, Iwamura H (1998) Inorg Chem 37: 2083–2085
30. Inoue K, Iwamura H (1994) J Chem Soc Chem Commun 2273–2274
31. Inoue K, Iwamura H (1995) Synth Met 71: 1791–1794
32. Inoue K, Hayamizu T, Iwamura H (1995) Mol Cryst Liq Cryst 273: 67–80
33. Inoue K, Hayamizu T, Iwamura H (1995) Chem Lett 745–746
34. Bonner JC, Fisher ME (1964) Phys Rev A135: 640–658
35. Coronado E, Drillon M, Georges R (1993) In: O'Connor CJ (ed). Research Frontiers in Magnetochemistry. Word Scientific, Singapore
36. Kahn O (1993) Molecular Magnetism. VCH
37. Scaling: 'Scaling factors employed here contain all the errors due to the sample purity, weighing the samples, calibration of SQUID susceptometer, etc. in addition to the deviation of the g factor from 2.00 as assumed in this study. If the last term is predominant, $f = g^2$ and, therefore, $f = 0.954$ corresponds to a g factor of 1.95 $(=2 \times \sqrt{0.954})$
38. Kumagai H, Hosokoshi Y, Markosyan AS, Inoue K (2000) New J Chem (in press)
39. Inoue K, Iwamura H (1994) J Am Chem Soc 116: 3137
40. Inoue K, Hayamizu T, Iwamura H, Hashizume D, Ohashi Y (1996) J Am Chem Soc 118: 1803–1804
41. Iwamura H, Inoue K, Koga N (1998) New J Chem 201–210
42. Iwahori F, Inoue K, Iwamura H (1999) J Am Chem Soc 121: 7264–7265
43. Gatteschi D, Laugier J, Rey P, Zanchini C (1987) Inorg Chem 26: 938–943
44. Laugier J, Rey P, Benelli C, Gatteschi D, Zanchini C (1986) J Am Chem Soc 108: 6931
45. Benelli C, Caneschi A, Gatteschi D, Laugier J, Rey P (1987) In: Delhaes P, Drillon M (eds), Organic and Inorganic Low Dimensional Crystalline Materials. Plenum Press, New York

46. Caneschi A, Gatteschi D, Laugier J, Rey P (1987) J Am Chem Soc 109: 2191
47. Caneschi A, Gatteschi D, Rey P, Sessoli R (1988) Inorg Chem 27: 1756
48. Caneschi A, Gatteschi D, Renard JP, Rey P, Sessoli R (1989) Inorg Chem 28: 2940
49. Benelli C, Caneschi A, Gatteschi D, Pardi L, Rey P (1989) Inorg Chem 28: 275
50. Caneschi A, Gatteschi D, Renard JP, Rey P, Sessoli R (1989) Inorg Chem 28: 2940
51. Caneschi A, Gatteschi D, Renard JP, Rey P, Sessoli R (1989) Inorg Chem 28: 3314
52. Caneschi A, Gatteschi D, Renard JP, Rey P, Sessoli R (1989) J Am Chem Soc 111: 785
53. Caneschi A, Gatteschi D, Renard JP, Rey P, Sessoli R (1989) Inorg Chem 28: 1976
54. Benelli C, Caneschi A, Gatteschi D, Pardi L, Rey P (1990) Inorg Chem 29: 4223
55. Cabello CI, Caneschi A, Carlin RK, Gatteschi D, Rey P, Sessoli R (1990) Inorg Chem 29: 2582
56. Caneschi A, Gatteschi D, Sessoli R, Cabello CI, Rey P, Barra AL, Brunel LC (1991) Inorg Chem 30: 1882
57. Iwamura H, Inoue K, Koga N, Hayamizu T (1996) In: Magnetism: A Supermoleculara Function. NATO ASI series, C484, Dordrecht
58. Inoue K, Iwamura H (1996) Mol Cryst Liq Cryst 286: 133
59. Inoue K, Iwamura H (1996) Proc Mater Res Soc 413: 313
60. Inoue K, Iwahori F, Markosyan AS, Iwamura H (2000) Coord Chem Rev (in press)
61. Iwahori F, Inoue K (to be published)
62. Markosyan AS, Iwamura H, Inoue K (1999) Mol Cryst Liq Cryst 334: 549–568
63. Markosyan AS, Hayamizu T, Iwamura H, Inoue K (1998) J Phys Condens Matter 10: 2323
64. Stanley HE (1971) Introduction to Phase Transitions and Critical Phenomena. Clarendon Press, London
65. Espie C, Laugier J, Ramasseul R, Rassat A, Rey P (1980) Nouv J Chim 4: 205
66. Espie JC, Ramasseul R, Rassat A, Rey P (1983) Bull Soc Chim Fr II 33
67. Laugier J, Ramasseul R, Rey P, Espie JC, Rassat A (1983) Nouv J Chim 7: 11
68. Benelli C, Gatteschi D, Zanchini C, Latour JM, Rey P (1986) Inorg Chem 25: 4242
69. Briere R, Giroud AM, Rassat A, Rey P (1980) Bull Soc Chim Fr II 147
70. Drillon M, Grand A, Rey P (1990) Inorg Chem 29: 771
71. Grand A, Rey P, Subra R (1983) Inorg Chem 22: 391
72. Ikorskii VN, Ovcharenko VI (1990) Zh Neorg Khim 35: 2093
73. Ovcharenko VI, Larionov SV, Mohosoeva VK, Volodarsky LB (1983) Zh Neorg Khim 28: 151; Chem Abstr 1983, 98: 118506w
74. Ovcharenko VI, Ikorskii VN, Vostrikova KE, Burdukov AB, Romanenko GV, Pervukhina NV, Podberezskaya NV (1990) Izv Sib Otd Akad Nauk SSSR Ser Khim Nauk 5: 100
75. Pervukhina NV, Ikorskii VN, Podberezskaya NV, Nikitin PS, Gel'man AB, Ovcharenko VI, Larionov SV, Bakakin VV (1986) Zh Strukt Khim 27: 61; Chem Abstr 1987, 106: 11395r
76. Caneschi A, Gatteschi D, Melandri MC, Rey P, Sessoli R (1990) Inorg Chem 29: 4228
77. Caneschi A, Gatteschi D, Sessoli R (1993) Inorg Chem 32: 4612
78. Markosyan AS, Hosokoshi Y, Inoue K (1999) Physics Letters A 261: 212–216
79. Tanaka M, Hosokoshi Y, Markosyan AS, Iwamura H, Inoue K (to be published)
80. McConnell HM (1963) J Chem Phys 39: 1910
81. Izuoka A, Murata S, Sugawara T, Iwamura H (1987) J Am Chem Soc 109: 2631

Magnetic Properties of Thiazyl Radicals

Jeremy M. Rawson[1], Fernando Palacio[2]

[1] Department of Chemistry, The University of Cambridge, Lensfield Road, Cambridge CB2 1EW, UK
    E-mail: jmr31@cam.ac.uk
[2] Instituto de Materiales de Aragon, CSIC-Universidad de Zaragoza, 30009 Zaragoza, Spain

A series of thiazyl radicals related to the trithiadiazolylium radical cation, $S_3N_2^{+\bullet}$ are described. In many instances the compounds exist as spin-paired singlets which are consequently diamagnetic. However, when the dimerisation process can be inhibited, these open shell molecules exhibit very strong exchange interactions, characterised by Weiss constants, θ, up to 10^2 K. Magneto-structural correlations show that, in the majority of instances, the magnetic exchange interactions are propagated via close intermolecular $S \cdots N$ and $S \cdots S$ contacts (typically in the region 3.1–3.7 Å). Examples of thiazyl radicals exhibiting magnetic ordering temperatures in excess of 50 K are described. In addition, the phenomenon of bistability (in which both open-shell monomeric and closed-shell dimeric forms are stable over the same temperature range) is discussed and examples of thiazyl radicals exhibiting bistability up to room temperature described.

Keywords: Thiazyl radical, Organic magnet, Molecular magnet, Ferromagnet, Antiferromagnet, Weak ferromagnet, Bistability

1	Introduction	94
2	Trithiadiazolylium Cations and Related Radicals	96
2.1	Electronic Structure.	97
2.1.1	Theoretical Calculations	97
2.1.2	EPR Spectra	98
2.2	Association of Thiazyl Radicals	99
2.2.1	Structural Studies	99
2.2.2	EPR Investigations of the Association Process	102
2.2.3	Theoretical Investigations into the Dimerisation Process	102
2.2.4	Implications of the Dimerisation Process on the Design of Thiazyl-Based Magnets	103
3	Magnetic Properties of 1,2,3,5-Dithiadiazolyl Radicals	104
3.1	Monomer or Dimer?	105
3.2	Radical Cations	105
3.3	Fluorinated Radicals	106
3.3.1	Magnetic Properties of p-BrC_6F_4CNSSN	107
3.3.2	Magnetic Properties of p-NCC_6F_4CNSSN.	108

3.3.2.1 Polymorphism 108
3.3.2.2 Electronic Structure 109
3.3.2.3 Magnetic Behaviour of α-p-NCC$_6$F$_4$CNSSN 110
3.3.2.4 Magnetic Behaviour of β-p-NCC$_6$F$_4$CNSSN 110

4   Magnetic Properties of 1,3,2-Dithiazolyl Radicals 113

4.1   1,3,2-Dithiazolyl Radicals without π-Stacking 113
4.1.1   Methylbenzo-1,3,2-Dithiazolyl, MBDTA 113
4.1.1.1 Electronic Structure 113
4.1.1.2 Structure 114
4.1.1.3 Magnetic Behaviour 114
4.1.2   Naphthalene-1,3,2-Dithiazolyl, NDTA 116
4.1.2.1 Crystal Structure 116
4.1.2.2 Electronic and Magnetic Behaviour 116
4.2   1,3,2-Dithiazolyl Radicals with π-Stacking 116
4.2.1   Quinoxaline-1,3,2-Dithiazolyl, QDTA 117
4.2.1.1 Electronic Structure 117
4.2.1.2 Crystal Structure 117
4.2.1.3 Magnetic Properties of QDTA 118
4.2.2   The Benzo-1,2:4,5-Bis(Dithiazolyl) Biradical, BBDTA 118
4.2.2.1 Electronic Structure 119
4.2.2.2 Crystal Structure 119
4.2.2.3 Electronic Properties 119
4.2.3   1,2,5-Thiadiazolo-Dithiazolyl-Pyrazine, TDP-DTA 120
4.2.3.1 Electronic Structure 120
4.2.3.2 Crystal Structure 120
4.2.3.3 Magnetic Properties of TDP-DTA 121
4.2.4   Trithiatriazapentalenyl Radical, TTTA 122
4.2.4.1 Electronic Structure 122
4.2.4.2 Crystal Structure 122
4.2.4.3 Magnetic Properties 123

5   Comment 125

5.1   Is It Possible to Make Room Temperature Organic Magnets? 125
5.2   Bistability in Organic Radicals 125

6   References 126

1
Introduction

In the last decade we have witnessed remarkable developments in the field of π-electron magnetism, specifically in the field of organic molecular magnets [1]. Fifteen or twenty years ago, the thought that a purely organic compound containing no metal ions could exhibit the magnetic behaviour which we typically associate with metals, their oxides, and ceramics was barely tenable. However, two reports [2, 3] in 1991 revolutionised the area of organic magnetism – suddenly the possibility that an organic material could exhibit long-range magnetic order became a reality. Work by Kinoshita and co-workers in Japan showed [2] that the open-shell radical p-nitrophenylnitronyl nitroxide, p-NPNN (Table 1) underwent a phase transition to a magnetically ordered state close to absolute zero (0.6 K), whilst Wudl's group [3] found that the fullerene charge-transfer salt, $C_{60} \cdot$ TDAE (Table 1) ordered as a ferromagnet at 16 K. Subsequently, there has been a continued flow of reports of organic radicals, especially nitroxide and nitronyl nitroxide radicals (Table 1) [4], that undergo magnetic ordering (ferromagnetism, antiferromagnetism, or weak ferromagnetism) at finite temperatures, although almost invariably below liquid helium temperature (4.2 K). In nearly all cases, the rather modest ordering temperatures can be ascribed to the weakness of the magnetic exchange interaction between neighbouring molecules. This arises because of the considerable steric protection required to inhibit dimerisation of the often reactive radical centre. In this article we review developments in thiazyl (sulfur-nitrogen) radical chemistry. In particular, we concentrate on a group of π radicals that are sterically unhindered and which provide close approach of radical centres and the potential for strong magnetic exchange interactions.

Table 1. Selected organic magnets and their transition temperatures (FM = ferromagnet, WFM = weak ferromagnet)

Compound	Ref.	Compound	Ref.
FM, 0.6K	[2]	FM, 0.7K	[4]
FM, 1.5K	[4]	FM, 0.2K	[4]
WFM, 36 K	[40]	FM, 16 K	[3]

2
Trithiadiazolylium Cations and Related Radicals

The trithiadiazolylium cation, $S_3N_2^{+\bullet}$, has been known for well over 100 years; it was first reported [5] by Demarcay in 1880, but it was not until the 1970s that its structures (both molecular and electronic) were elucidated [6]. A number of structures have been determined in the solid state [7] and all have been found to be dimeric with the two radicals associated via a long S···S contact (ca. 2.9–3.0 Å) between rings. This S···S contact is significantly longer than a conventional S—S bond (2.08 Å) [8], although somewhat shorter than the sum of the van der Waals radii perpendicular to the heterocyclic ring (4.06 Å) [9]. EPR studies [6] combined with theoretical calculations [6] showed that the unpaired electron resides in a π^* orbital of a_2 symmetry. The dimerisation process can therefore be considered as a four centre, two electron π^*-π^* bonding interaction between S atoms. A qualitative MO diagram of this interaction, together with the structure of $[S_6N_4]Cl_2$ is shown in Fig. 1.

A number of related radicals derived from the $S_3N_2^{+\bullet}$ cation can be prepared by replacing S^+ by the isoelectronic groups R—C or N (see Scheme 1). These radicals will be described in more detail in later sections. More recently, there has been considerable success in developing selenium analogues of these radicals [10, 11]. In this article we will focus on the family of S/N radicals 1–5 depicted in Scheme 1.

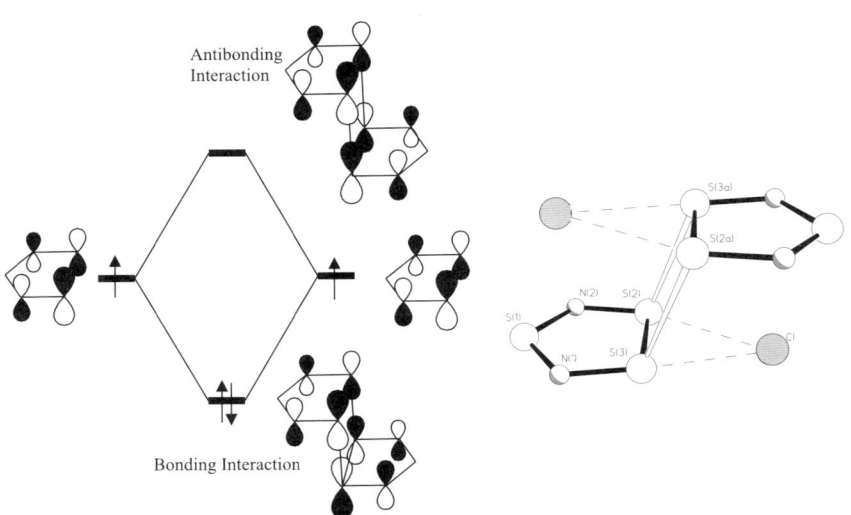

Fig. 1. Qualitative MO diagram for the spin-paired association of two $S_3N_2^{+\bullet}$ radicals, and structure of $[S_3N_2]_2Cl_2$

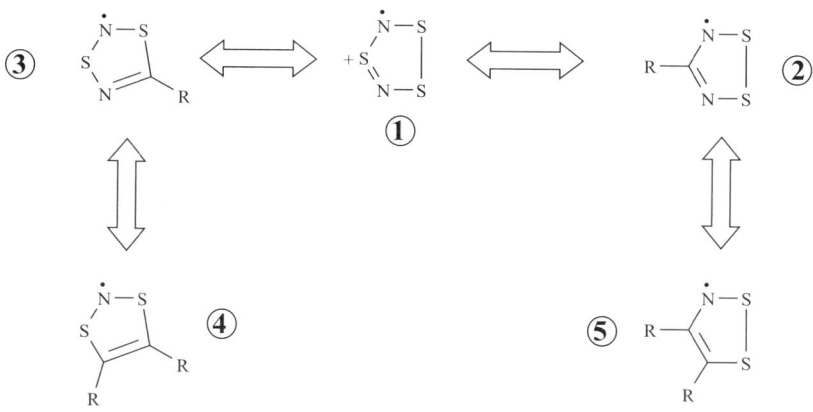

Scheme 1. A family of heterocyclic radicals derived from $S_3N_2^{+\bullet}$

2.1
Electronic Structure

The electronic structures of these π radicals have been carefully probed by a combination of experimental (EPR) and theoretical calculations by a number of research groups and the results are summarised here.

2.1.1
Theoretical Calculations

Theoretical calculations on $S_3N_2^{+\bullet}$ and its derivatives have been made at a variety of levels, from Hückel theory [12] to *ab initio* methods [13] to Density Functional Methods [14]. The π manifolds of these thiazyl radicals are closely related to those of the cyclopentadienyl anion, Cp^- [12a]. However, the degeneracies of the π system observed in Cp^- are lifted due to the lower symmetry and mismatch in energies of the constituent atoms (C/N/S). In all cases the symmetry of the singly occupied molecular orbitals correspond closely to either π_4 or π_5 of Cp^-. The π manifold of Cp^- is shown in Fig. 2, along with the SOMOs of 1–4. Both 1 and 2 have SOMOs of a_2 symmetry under the C_{2v} point group corresponding to π_4 of Cp^-. Radicals 3 and 4 have orbitals corresponding to π_5 of Cp^-. Notably in derivatives of 1–4 there is little delocalisation of unpaired spin density away from the heterocyclic ring. This arises because of the difference in the energies of the π-manifolds of the heterocyclic ring and the aryl substituent. Calculations on the open shell radicals also indicate significant polarity of the S—N bond, in the sense $S^{\delta+}$—$N^{\delta-}$. The localisation of the spin density and the polarity of the heterocyclic ring have important consequences for the magnetic behaviour of thiazyl radicals (see Sect. 2.2.4).

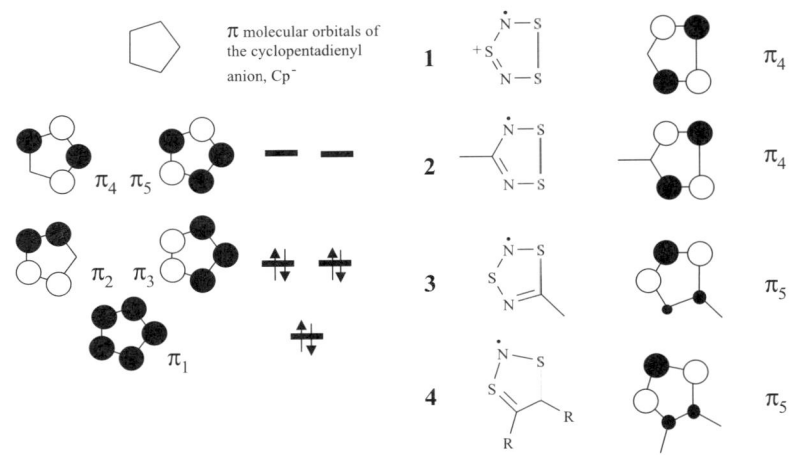

Fig. 2. The π-orbitals of Cp⁻ and the singly occupied molecular orbitals of **1-4**, corresponding to either π_4 or π_5 of the Cp⁻ ring

2.1.2
EPR Spectra

EPR studies on **1-5** all show extensive delocalisation around the heterocyclic ring. Their EPR spectra are characterised by isotropic g-values typically in the region 2.005–2.010 with the deviation from the free electron value caused in large part by the larger spin-orbit coupling associated with S. Indeed, where selenium analogues are known, this gives rise to even greater shifts in the g-factor [10b, 15]. In addition, frozen glass spectra indicate considerable anisotropy in the g-value. These radicals exhibit a rhombic EPR spectrum with one set of large hyperfine coupling constants to N in the high-field component. This is entirely consistent with a π-based radical in which the π-type nature of the orbital means that significant hyperfine coupling is seen in only the direction perpendicular to the heterocyclic ring plane. This is clearly evidenced in the EPR spectrum of PhCNSSN (Fig. 3). Some typical EPR data for derivatives of **1-5** are given in Table 2. There is typically a good correspondence between the calculated unpaired spin density distributions (both *s* and *p* contributions) from EPR [10b, 15] and those predicted from theoretical calculations [13, 14].

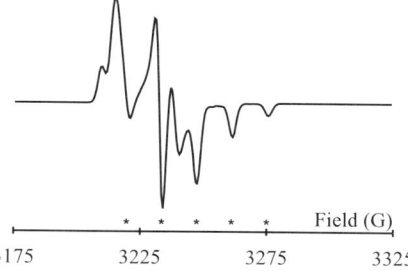

Fig. 3. Frozen glass X-band EPR spectrum of PhCNSSN at 77 K. Note the rhombic pattern and large hyperfine coupling to N (marked *) to just the g_1 component, characteristic of these π-based radicals

2.2
Association of Thiazyl Radicals

2.2.1
Structural Studies

A number of structural studies on **1** have shown a strong tendency for these π-radicals to associate through an out-of-plane π*-π* interaction (described previously and shown in Fig. 1). In derivatives of **1**, the S···S separation between monomers is typically 2.9–3.0 Å [7], somewhat less than the sum of the van der Waals radii perpendicular to the ring.

Whilst only one mode of association (*trans*-antarafacial) has been observed for **1**, the modes of association of the dithiadiazolyl radical **2** are considerably more diverse. The mode of association has been clearly linked to the bulkiness of the R substituent. When R is not sterically demanding then the common mode [16] of association is a *cis*-cofacial geometry (Fig. 4). For more sterically demanding groups (e.g., R = Me, CF$_3$, Me$_2$N, adamantyl) [13a, 13c, 17] then a twisted geometry is obtained (Fig. 4b). More recently, phenyl substituents containing a strong dipole across the molecule have also been found to induce twisting to maximise dipolar interactions [18]. In one reported instance a *trans*-antarafacial geometry has been observed [16b] (Fig. 4c). A fourth *trans*-cofacial mode of association has also been reported (Fig. 4d) [19]. With the exception of this latter mode, all these associative conformations exhibit one or more intradimer S···S contacts, typically in the range 2.9–3.1 Å [16–18]. In the case of the *trans*-cofacial mode depicted in Fig. 4d, the S···N contacts are in the range 3.12–3.35 Å, similar to the S···N contacts in the twisted configuration (3.18–3.24 Å) [13a]. In a few instances monomeric or intermediate structures have been observed, and these will be described later.

In all these modes of association, a bonding interaction between SOMOs can be clearly identified. Theoretical calculations by Mews and co-workers [13a] indicated that the calculated difference in energy between the modes of

Table 2. EPR data for derivatives of 1–5 [hyperfine coupling constants are quoted in G]

Compound	g_x	a_x^N	$a_x^{N'}$	g_y	a_y^N	$a_y^{N'}$	g_z	a_z^N	$a_z^{N'}$	g_{iso}	a_{iso}^N	$a_{iso}^{N'}$	Ref.
1	2.0013	9.18	–	2.0062	~0	–	2.0250	~0	–	2.01112	3.19	–	[6c]
2	2.0021	14.10	–	2.0078	1.07	–	2.0218	0.35	–	2.0102	5.19	–	[6c]
3	2.0012	27.85	2.46	2.0012	2.5	~0	2.0120	2.5	~0.1	2.0048	10.94	0.525	[59b]
4	2.0016	27.1	–	2.0055	2.5	–	2.0126	2.5	–	2.00707	10.66	–	[59b]
5	*	*	–	*	*	–	*	*	–	2.0081	8.13	–	[a]

[a] Harrison S, Pilkington, RS, Sutcliffe LH (1984) *J Chem Soc Faraday Trans I* 80: 669.

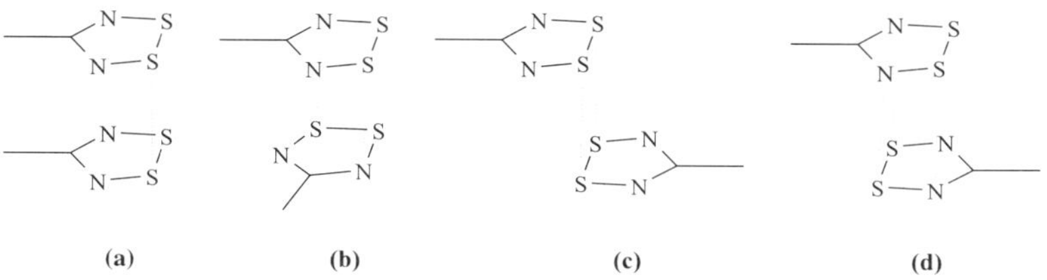

Fig. 4. Modes of association of dithiadiazolyl radicals: (a) *cis*-cofacial, (b) twisted, (c) *trans*-antarafacial, (d) *trans*-cofacial

association was small (~5 kJ mol^{-1}) and the overall geometry is likely to be dictated, in a large part, by other packing forces.

Only one structure of a 1,3,2,4-dithiadiazolyl radical, 3, has been reported; the diradical *p*-SNSNCC$_6$H$_4$CNSNS also associates in the solid state in a *trans*-cofacial fashion, held together by S···S (3.21 Å) and S···N (3.35 Å) contacts [20]. Attempts to prepare other derivatives have been thwarted by the propensity of derivatives of 3 to rearrange to 2 either in solution [21] or in the solid state [22]. More recently, Passmore prepared crystals of PhCNSNS, although these crystals decomposed in the X-ray beam [23].

A number of structures of the 1,3,2-dithiazolyl ring, 4, are known and two common modes of association have been observed, either *cis*-cofacial [24] or *trans*-cofacial [25] (Fig. 5). In both cases the dominant interaction would appear to be intradimer S···S contacts (3.14–3.18 Å). However, for many derivatives of 4, a far greater diversity of structure is observed, with many compounds exhibiting stacked structures with intermolecular separations significantly greater than 3.5 Å. These will be discussed later.

Only one derivative of 5 has been structurally characterised (5, R^1 = C_6F_5, R^2 = Cl) [26]. It too associates, in a twisted configuration (to avoid steric hindrance between perfluoroaryl groups) with an intradimer S···S separation of 3.30 Å.

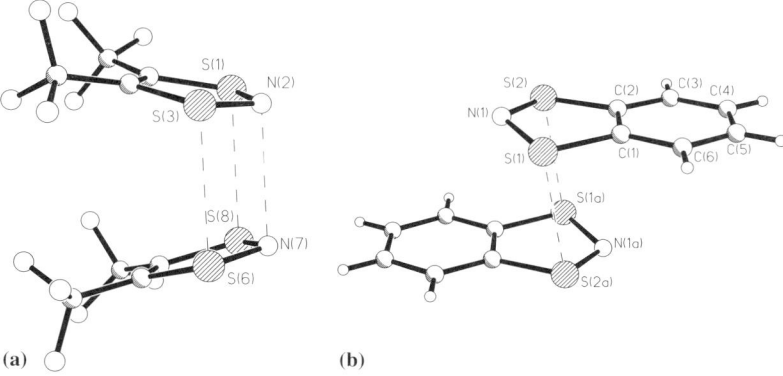

Fig. 5. Crystal structures of (a) [F$_3$CCSNSCCF$_3$]$^•$ and (b) C$_6$H$_4$S$_2$N$^•$

2.2.2
EPR Investigations of the Association Process

In solution, these dimers typically dissociate into monomers but a monomer-dimer equilibrium exists (see Eq. 1). The equilibria can be monitored by UV-visible spectroscopy [27] or by measuring the intensity of the EPR spectrum [6c, 21b, c, 28]. Careful EPR studies have allowed the thermodynamics of the equilibrium to be determined in some cases. Some data are given in Table 3. Its importance is that it provides an estimate of the dimerisation energy; if these thiazyl radicals are to be used in the design of magnetic materials, then this dimerisation energy must be overcome. As can be seen from Table 3, the dimerisation energy for **1** is ca. 45 kJ mol^{-1} and invariably the solid state structures of derivatives of **1** are dimeric. In comparison, replacement of one S by C, leads to a reduction in dimerisation energy such that ΔH_{dim} for **2** and **3** are ca. 35 and 20 kJ mol^{-1}, respectively. Whilst the majority of derivatives of **2** are also dimeric in the solid state, a few are monomeric and will be discussed in more detail later. For **4**, replacement of N by C leads to a further reduction in ΔH_{dim} (~ 0 kJ mol^{-1}) and a large number of these radicals are found to be monomeric in the solid state.

2.2.3
Theoretical Investigations into the Dimerisation Process

A number of theoretical studies have been made on the dimerisation of **1**, **2**, **3**, and **4**. All conclude that there is a π^*-π^* interaction between singly occupied molecular orbitals (SOMOs) which leads to a net bonding interaction and a closed shell singlet ground state [13a, 25, 29]. A variety of modes of association is possible that give rise to efficient overlap of π orbitals, and all require the rings to be essentially eclipsed, or rotated by 90°

Table 3.

1		2		3		4	
S-N=S-N=S (structure)	-47^a	Ph-C (structure)	-37^a	F$_3$C-C (structure)	-19^c	^{t}Bu-C (structure)	$\sim 0^d$
		Ph-C (structure)	-37^b			F$_3$C, F$_3$C (structure)	$\sim 0^d$
		(benzo structure)	-31^b				

^{a} Ref. [6c];
^{b} Ref. [21b];
^{c} Ref. [21c];
^{d} Ref. [28].

or 180° with respect to one another and these are all observed structurally. A more quantitative analysis of the association has proved problematic, and we highlight this by a recent, detailed *ab initio* study [13d] of the dimerisation process by Oakley and co-workers at the University of Guelph. They examined how the theoretical model of the dimerisation process changed as a function of the basis set employed, paying particular attention to the refined geometry and energy of dimerisation. Initial findings showed that valence minimal basis sets produced significant distortions of the dimeric structures and appeared to maximise N···N rather than S···S inter-ring bonding interactions. Additional *d*-functions (3-21G*) were successfully employed and produced much better agreement with the observed crystallographic parameters. However, the calculated energies of dimerisation proved to be particularly sensitive to the theoretical method used. At low orders of perturbation theory (MP2) the dimerisation energy was found to be 176 kJ mol^{-1}, but at third (MP3) and fourth (MP4SDQ) order these energies reduced successively to 27 and then 17 kJ mol^{-1} (cf. experimental values of ca. 35 kJ mol^{-1} in Table 3). This variation in enthalpy of dimerisation with basis set provides a substantial theoretical problem and calculated enthalpies of association should not be taken quantitatively. It should, of course, be remembered that these calculations are gas-phase and the energies considered are small. In such circumstances, both solvent and certainly lattice effects may provide a significant perturbation. Nevertheless, a set of comprehensive studies using the same basis set on derivatives of 1–5 might, however, provide a useful correlation (e.g., between calculated and observed values of ΔH_{dim}) for future work.

2.2.4
Implications of the Dimerisation Process on the Design of Thiazyl-Based Magnets

These thiazyl radicals exhibit a tendency to associate both in solution and in the solid state through a spin-paired dimerisation process, generating a closed shell singlet ground state. If they are to be utilised in the design of molecular magnets, then this dimerisation process must be overcome. The smaller the dimerisation enthalpy is, the more favourable is the possibility of forming paramagnetic structures in the solid state. This is borne out by a simple correlation of the structural data available on the known derivatives of 1–5 and the estimates of the dimerisation energy for derivatives of 1–5. Evidently, severe steric and/or electronic demands must be placed on radicals such as 1–3 if the dimerisation process is to be inhibited. In contrast, derivatives of 4 are likely to be monomeric in many instances.

There are several further important factors which should be considered here. Firstly, the unpaired spin density in these thiazyl radicals is, to a large degree, localised on the heterocyclic ring. The consequence is that we can vary the substituents to modify the molecular structure, without significantly changing the electronic structure. Secondly, the thiazyl bonds are extremely polar and there is a great tendency for these radicals to pack in such a way as to maximise the intermolecular $S^{\delta+}···N^{\delta-}$ interactions. This leads to close approach of regions of unpaired spin density. Provided that the out-of-plane SOMO-SOMO dimerisation process can be inhibited, the close approach of radical centres should lead to strong exchange interactions. Since the heterocyclic N and S atoms both bear considerable unpaired spin density the S···N interactions are likely to propagate antiferromagnetic exchange interactions (on the basis of McConnell theory [30]). In the sections which follow we examine the magnetic properties of derivatives of 2 and 4.

3
Magnetic Properties of 1,2,3,5-Dithiadiazolyl Radicals

In this section we examine how the dimerisation process (ca. 35 kJ mol^{-1}) can be overcome through a combination of electronic and steric factors. Since the spin density can be assumed to be localised on the heterocyclic ring, then particular attention will be paid to the intermolecular interactions between dithiadiazolyl rings since these are likely to propagate the magnetic exchange pathway through the solid. Correlations between magnetic behaviour and solid state structure will be made.

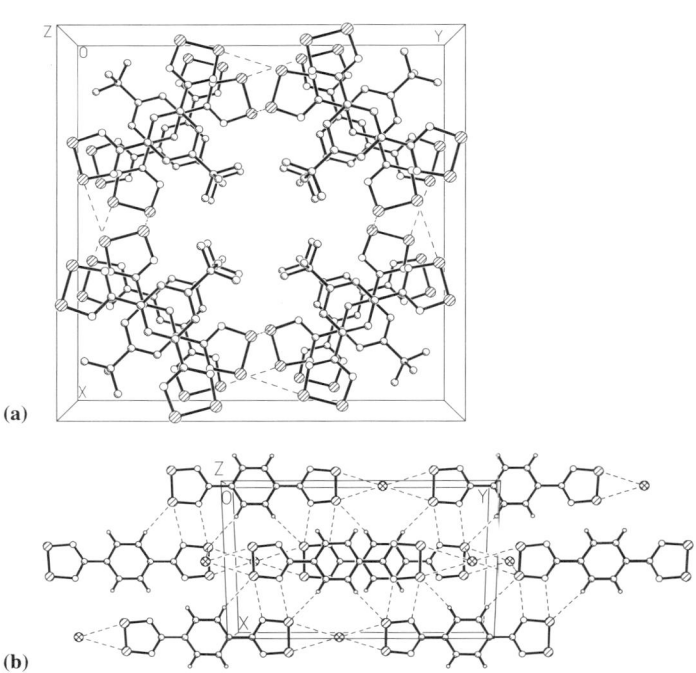

Fig. 6. Crystal structures of (a) **2a** and (b) **2d**

3.1
Monomer or Dimer?

The majority of dithiadiazolyl radicals associate as spin-paired dimers in the solid state in which the intermolecular separation is in the region 2.9–3.1 Å. However, one example has been reported in which the intermolecular separation is larger. The bis-radical **2a** also has a long intermolecular stacking distance with the shortest contact along the stacking direction (see Fig. 6a) at 3.48 Å [31]. The long intermolecular separation is still within the sum of the van der Waals radii of S (4.0 Å). Radical **2a** exhibits some paramagnetism at room temperature (equivalent to ca. 30% unpaired spins), but its susceptibility drops rapidly such that it is rendered diamagnetic by 200 K. A second radical **2b** was reported to exhibit a stacking motif with a regular lattice spacing of 3.54 Å, coincident with the crystallographic c-axis. [18a] However, **2b** appeared to be diamagnetic throughout the temperature range studied, despite the longer intermolecular separation! A recent re-investigation [18b] of the crystal structure of **2b** showed that a super-cell was present in which there is regular packing of dimers along the π-stacking direction. As with other dimeric dithiadiazolyl radicals, the short intra-dimer contacts render the material diamagnetic.

3.2
Radical Cations

A number of derivatives of **2** have been reported in which the radical bears a cationic substituent. Under these circumstances the ionic packing forces may dominate the packing and overcome the dimerisation energy. In 1993, Banister and co-workers reported [32] the radical cation **2c**. Whilst there was no crystallographic data, the magnetic moment at room temperature was 1.8μ_B and it exhibited Curie–Weiss behaviour with $\theta = -65$ K. Such large values for the Weiss constant will appear as a common theme in this area of chemistry and are consistent with strong communication between radical centres. More recently, the radical cation **2d** was isolated [33]. Its structure is shown in Fig. 6. This compound too has a room temperature moment consistent with one unpaired electron (ca. 1.3 μ_B). On cooling the susceptibility passes through a broad maximum around 275 K consistent with the onset of short-range antiferromagnetic order. It then decreases slowly to ca. 50 K before undergoing an abrupt transition to a diamagnetic state at 15 K. This has been ascribed to a structural phase transition to a diamagnetic (dimeric) state [33].

3.3
Fluorinated Radicals

In recent years our own research has developed a series of perfluoroaromatic derivatives of **2**. In these derivatives the electrostatic repulsions between the *ortho*-fluorine and the heterocyclic ring N leads to a large twist angle between the two rings (30–60°), thereby making the substituent more sterically demanding. In addition, fine control of the electronic effects of the substituent have been achieved. For a series of derivatives **2e–k**, a correlation is observed (Fig. 7) between both the electron-withdrawing/electron-releasing nature of the *para*-substituent (Hammett parameter) and the ease of reduction, i.e., radical stability. In addition the structures which appear to bear strongly electron-withdrawing groups have been found to be monomeric.

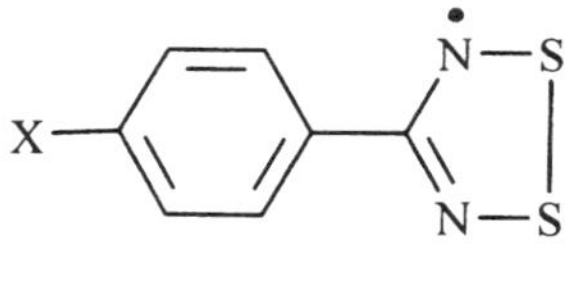

2e X = F
2f X = Cl
2g X = Br
2h X = MeO
2i X = Me
2j X = NO$_2$
2k X = H

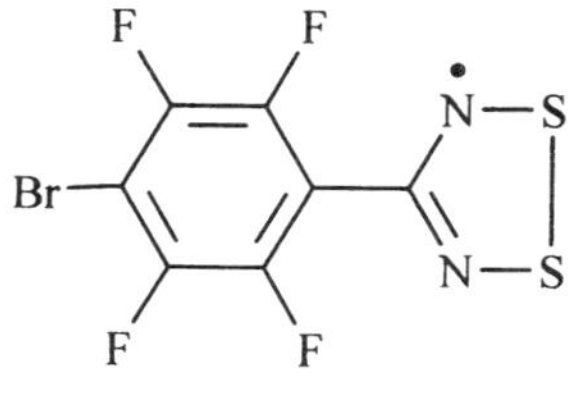

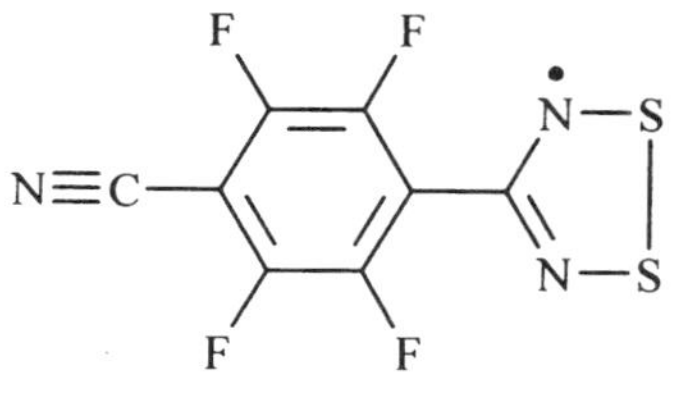

2l

2m

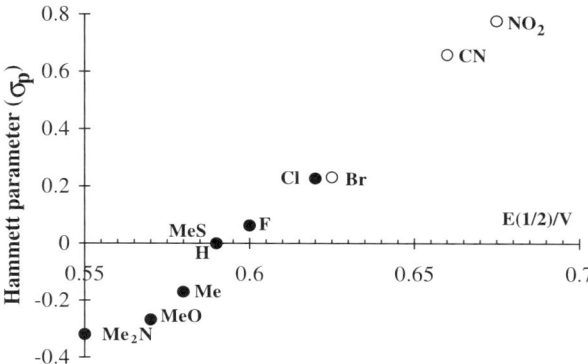

Fig. 7. Correlation between the Hammett parameter, σ_p of the *para*-substituent and the reduction potential of the corresponding $[p\text{-}XC_6F_4CNSSN]^+$ ion. Solid state structures have been determined for most derivatives $p\text{-}XC_6F_4CNSSN$. Those marked (●) are dimeric, those marked (○) are monomeric

3.3.1
Magnetic Properties of $p\text{-}BrC_6F_4CNSSN$

The electronic structure of $p\text{-}BrC_6F_4CNSSN$, **2l**, has been investigated [34] by EPR studies (X-band solution EPR and ENDOR) and DFT calculations. Solution EPR spectra clearly resolve the nitrogen hyperfine interaction, and the much smaller F and Br couplings can be detected in the ENDOR experiments. An analysis of the spin density distribution is in excellent agreement with DFT calculations. The resultant spin density distribution is shown in Fig. 8 and clearly shows that essentially all the unpaired spin density is localised on the N_2S_2 fragment of the heterocyclic ring. The dominant magnetic exchange is therefore likely to be via close intermolecular contacts between heterocyclic rings.

The structure of **2l** is shown in Fig. 9. In **2l** the intermolecular contacts along the stacking direction (crystallographic *a* axis) are in the range 3.67–4.00 Å, although the twist between the rings precludes efficient orbital overlap [35]. Instead molecules link together via electrostatic $S^{\delta+}\cdots N^{\delta-}$ interactions (S$\cdots$N = 3.18 Å) along the crystallographic *c* axis. These chains are isolated from each other by ca. 10.2 Å, so that the magnetic exchange pathway can, at best, be considered to be two-dimensional. The room temperature magnetic moment (1.5 μ_B) is close to that expected for an $S = 1/2$ ion. Above 60 K the compound exhibits Curie–Weiss behaviour ($\theta = -27 \pm 1$ K). Using a mean field approximation, the compound would be expected to undergo a paramagnetic-antiferromagnetic phase transition at 27 K. However, no evidence of long-range order is observed down to 1.8 K, consistent with the low-dimensional nature of the structure. Attempts to model the behaviour using the Bonner-Fischer model for a Heisenberg chain of $S = 1/2$ ions with a constant exchange term (*J*) proved unsuccessful since the observed data did not exhibit a maximum in the susceptibility. An excellent fit was obtained

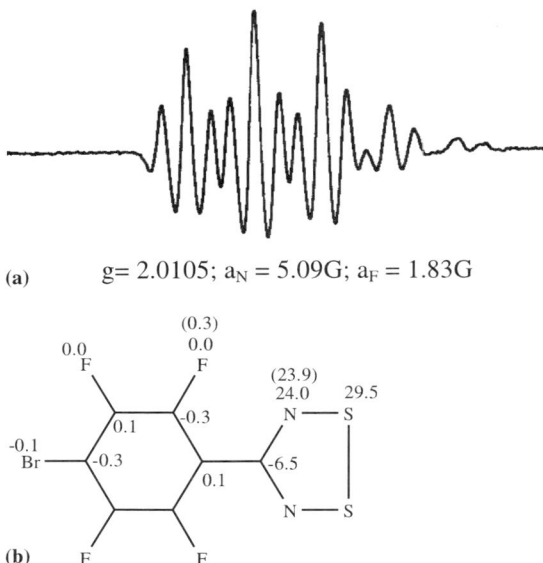

Fig. 8. (a) Solution EPR spectrum of **2l** and (b) theoretical spin density distributions. Estimated values from experimental measurements are given in parentheses

(Fig. 9b), however, if *J* was allowed to be temperature-dependent, such that the magnetic exchange between radical centres decreased on cooling. Further experiments will be needed to determine whether there is any physical basis for this phenomenon, e.g., changes in intermolecular contact, etc.

3.3.2
Magnetic Properties of p-NCC$_6$F$_4$CNSSN

3.3.2.1
Polymorphism

A recurrent theme in thiazyl chemistry and indeed in the area of organic molecular magnetism is the propensity for compounds to crystallise in two or more space groups, i.e., polymorphism. Since the solid state structure determines the strength and nature of the magnetic exchange interaction, then the magnetic behaviour of different polymorphs cannot be expected to be the same. This phenomenological problem has been reviewed recently [36].

In the area of dithiadiazolyl and diselenadiazolyl chemistry, at least four compounds have been reported to be polymorphic [16b, 37–39] in recent years. The dithiadiazolyl radical p-NCC$_6$F$_4$CNSSN, **2m**, provides a classic example of this phenomenon. Radical **2m** crystallises in either of two space groups [38, 40], depending on the sublimation conditions: sublimation under high vacuum at low temperatures leads to the formation of the kinetic phase, α-**2m** (triclinic, P-1) [38] whereas slow sublimation at more elevated

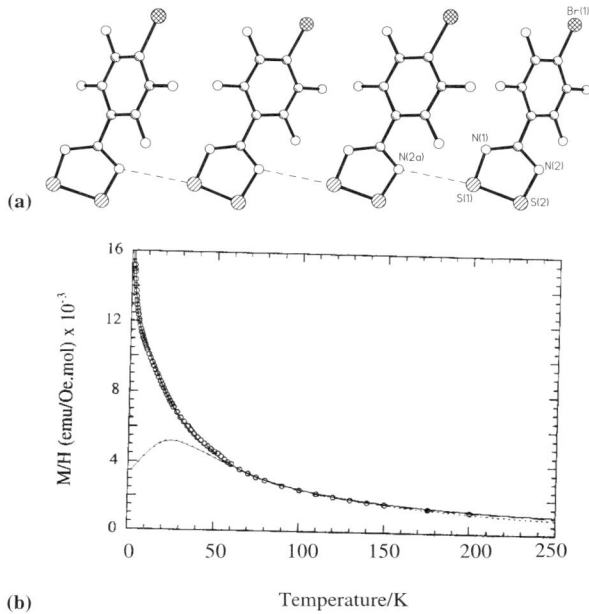

Fig. 9. (a) View of **2l**, viewed down the crystallographic a axis, illustrating the close in-plane S⋯N interactions (3.18 Å) and (b) Plot of molar susceptibility of **2l** as a function of temperature. The fit to a Bonner-Fischer model with temperature-dependent J [$J = -(0.31T + 8.2)$] is shown as a solid line in the Figure. The dashed line represents the best curve-fit to the high temperature data with a fixed J term

temperatures leads to the thermodynamic phase, **β-2m** (orthorhombic, $Fdd2$) [40]. The structures of the α- and β-phases are shown in Fig. 10. There are some marked similarities in the two phases: both phases are monomeric with molecules linked together into chains via electrostatic $CN^{\delta-}\cdots S^{\delta+}$ interactions. Studies on thin films of **2m** indicate [41] that this CN⋯S interaction plays a role in directing crystal growth. Indeed in **β-2m**, the crystals grow as long needles with the CN⋯S interaction (along the crystallographic c axis) coincident with the needle axis of the crystal [42]. However in the α-phase, these chains of radicals crystallise in an antiparallel fashion with inter-chain interactions between the S and the cyano-N. In the β-phase, the interchain interactions are between the S and the heterocyclic N and generate coparallel alignment of the chains. The lack of an inversion centre in **β-2m** has important ramifications for its magnetic behaviour.

3.3.2.2
Electronic Structure

The electronic structure of **2m** has been probed by solution EPR studies (X-band and ENDOR) and also through DFT calculations [34]. The electronic

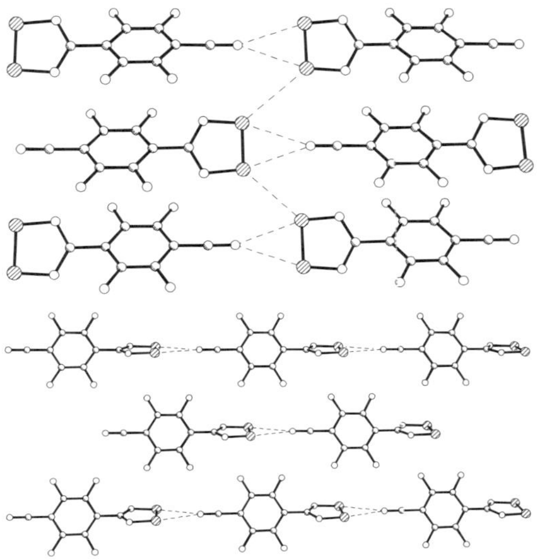

Fig. 10. Crystal structure of (a) α-p-NCC$_6$F$_4$CNSSN and (b) β-p-NCC$_6$F$_4$CNSSN

structure of **2m** is almost identical to that of **2l** and there is good agreement between theoretical and experimental spin density distributions.

3.3.2.3
Magnetic Behaviour of α-p-NCC$_6$F$_4$CNSSN

The α-phase of **2m** exhibits a room temperature susceptibility of ca. 1.6 μ_B consistent with an $S = 1/2$ ion [38]. It obeys Curie-Weiss behaviour above 175 K with $\theta = -25$ K. On cooling further, deviation from Curie–Weiss paramagnetism is observed, with the sample passing through a minimum in χ^{-1} at about 8 K; a little lower than that expected from the mean field theory approximation. The problematic synthesis of the α-phase (the thermodynamically stable β-phase grows preferentially) has precluded further study of its low temperature behaviour. These studies should confirm whether the compound undergoes long-range order at 8 K or whether it is a short range phenomenon. An analysis of the magnetic exchange pathways [42] indicates a number of intermolecular S···S contacts (3.60–3.67 Å) which should propagate the magnetic exchange in at least two dimensions.

3.3.2.4
Magnetic Behaviour of β-p-NCC$_6$F$_4$CNSSN

The β-phase of **2m** also exhibits a room temperature susceptibility of 1.6 μ_B. It also obeys Curie–Weiss behaviour above 120 K, but with a much larger Weiss constant, $\theta = -102$ K indicative of a strongly antiferromagnetically coupled regime [40]. As the temperature decrease further, χ passes through a broad maximum at 60 K, indicative of the onset of short-range order. A fit of the high temperature data to a Bonner-Fischer expression for an $S = 1/2$ Heisenberg linear chain yields an intrachain exchange term of -83 K, but also a large interchain term [$zJ_{inter}/J_{intra} = \sim 0.1$]. The large magnitude of the interchain interactions means that β-2m is not a good one-dimensional system [40]. An analysis of the intermolecular contacts shows that each molecule forms four close heterocyclic-N$\cdots$S contacts to other molecules in a distorted tetrahedral geometry. Propagation of these magnetic exchange interactions through the lattice lead to a distorted diamond-lattice for the magnetic exchange interactions [43]. Inclusion of the related radical p-NCC$_6$H$_4$CNSSN, which has a very similar spin density distribution, to 2m into the host-lattice of PHTP has confirmed that there is negligible exchange via the CN$\cdots$S interaction [43].

On cooling below 36 K, the susceptibility of β-2m becomes field dependent and an out-of-phase component in the ac susceptibility is observed (Fig. 11), consistent with the onset of long range order [44]. The field-dependent magnetic behaviour of β-2m below 36 K can be fitted to Eq. 1,

$$M = \chi H + M_S \qquad (1)$$

The small residual spontaneous magnetisation observed in zero-field can be explained in terms of spin-canting of the antiferromagnetic lattice leading to a small spontaneous magnetic moment. Hysteresis can be observed (Fig. 11) at low temperature with a small coercive field and remnant magnetisation.

Canted antiferromagnetism (weak ferromagnetism) is favoured in low-dimensional systems and results from antiferromagnetic alignment of spins on two magnetic sublattices which are equivalent in number but not exactly antiparallel [45]. The sublattices are canted in the absence of a magnetic field provided that the total symmetry is unchanged and that this leads to a net magnetisation. Spin-canting cannot be observed in compounds whose spins

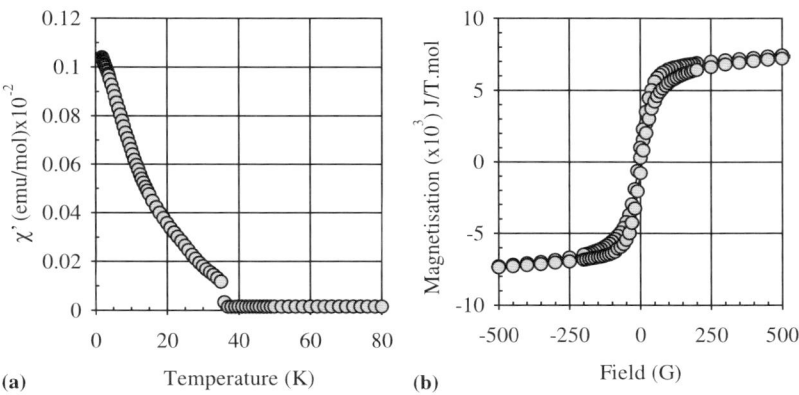

Fig. 11. (a) ac susceptibility of β-p-NCC$_6$F$_4$CNSSN below 80 K and (b) hysteresis of β-p-NCC$_6$F$_4$CNSSN recorded at 1.8 K

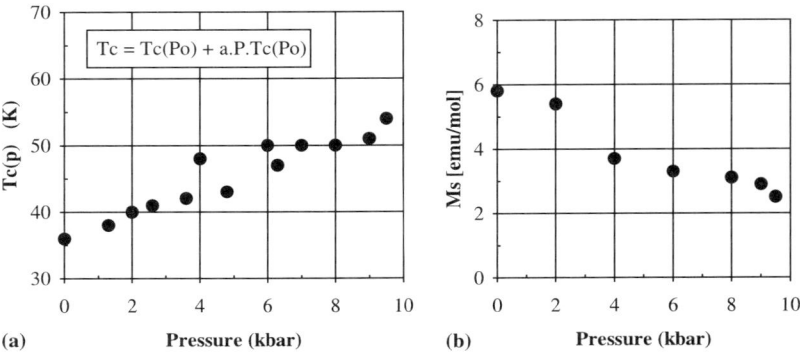

Fig. 12. (a) The variation of Tc of $\beta\text{-}p\text{-NCC}_6\text{F}_4\text{CNSSN}$ with applied pressure, and (b) the variation of the spontaneous magnetisation, Ms, with applied pressure

(on different sublattices) are related through an inversion centre (e.g., α-2m) but can be achieved in non-centrosymmetric space groups, such as β-2m. Canting arises through the Dzyaloshinskii-Moriya interaction which minimises the exchange energy when the spins are orthogonal. The magnitude of the canting angle is proportional to the anisotropy in the g-tensor (Eq. 2) and, as a consequence, can be estimated from EPR studies. The canting angles determined from the saturation magnetisation extrapolated to 0 K (1.5×10^{-3} μ_B), and predicted from EPR studies are in good agreement, cf. $0.085°$ and $0.117°$ [46].

$$\theta = 1/2 \arctan(|d/J|) \text{ where } |d/J| \sim (g - g_e)/g \tag{2}$$

The onset of long-range magnetic order has been probed using specific heat capacity studies [44], powder [44] and single crystal [47] neutron diffraction, and muon-spin relaxation [48]. Single crystal symmetry precludes c-axis spin-alignment for a canted structure and the resultant magnetic moment must align parallel with the crystallographic a or b axes [44]. A careful analysis of the single crystal, powder, and muon data is in agreement that the moments are oriented parallel to b, with a small canted moment along the crystallographic a axis.

The magnetic behaviour of β-2m is truly exceptional. The strength of the magnetic exchange interaction, propagated through the close S$\cdots$N contacts, gives rise to a T_C of 36 K – the first example of an open shell radical to exhibit a T_C above liquid helium temperature. Recent studies have shown [49] that the application of pressure leads to a further increase in T_C. This presumably arises out of a shortening of the intermolecular S$\cdots$N contacts thereby enhancing the magnetic exchange term, J. The increase in T_C appears linear with the applied pressure and fits the empirical formula,

$$T_C(P) = T_C(P_o) + aPT_C(P_o) \tag{3}$$

It should be noted that the ordering temperature under a pressure P is dependent not only on the parameter a but also on Tc(P = 0). Since Tc(P = 0) is large, then the enhancement in Tc is marked, reaching 50 K at 7 kbar (Fig. 12). Notably, however, the increase in the magnitude of the exchange term J also leads to a reduction in $|d/J|$. Necessarily (according to Eq. 2), this also leads to a reduction in canting angle and the value of the spontaneous magnetisation also decreases (Fig. 12) [49].

4
Magnetic Properties of 1,3,2-Dithiazolyl Radicals

In comparison to the dithiadiazolyl radicals, **2**, described above, the dithiazolyl radicals, **4**, exhibit dimerisation energies close to 0 kJ mol^{-1}. As a consequence, many are found to be monomeric in the solid state. We will separate the magnetic behaviour of this class of radical into two groups based on some structural motifs: those which contain π-stacked structures in the solid state and those which do not. We will concentrate in the first instance on those in which π-stacking is not observed.

4a   **4b**

4.1
1,3,2-Dithiazolyl Radicals without π-Stacking

4.1.1
Methylbenzo-1,3,2-Dithiazolyl, MBDTA

The MBDTA radical was first reported by Wolmershauser et al. in 1984 [50]. They reported that both the benzo-derivative BDTA, **4a** and the tolyl derivative, MBDTA, **4b**, were paramagnetic at room temperature. Subsequent studies by Passmore showed that **4a** was in fact a dimer [25]. However, our own structural studies on MBDTA indicate that it is indeed monomeric [51].

4.1.1.1
Electronic Structure

EPR studies on **4b** exhibit a triplet pattern (g = 2.003, a_N = 11.4G) consistent with localisation of the unpaired spin density on the heterocyclic ring [51]. The electronic effect of the methyl group is likely to be minimal and the spin density distribution can be expected to closely parallel that of **4a** calculated by Passmore and co-workers with over 75% of the unpaired spin density residing on the C_2S_2N ring. Whilst there is evidently more π-delocalisation of the unpaired spin density in derivatives of **4**, (cf. **2**) it is by no means extensive.

Thus, heterocyclic contacts are likely to continue to play a dominant role in determining the magnetic properties of these radicals. (The SOMO of **4b** is shown in Fig. 13).

4.1.1.2
Structure

Radical **4b** crystallises in the orthorhombic space group *Pb*ca. The asymmetric unit of **4b** contains a single molecule of unexceptional geometry, with the fused ring essentially planar. Each molecule has six nearest neighbours (Fig. 13) with intermolecular contacts in the range 3.71–3.82 Å. This set of contacts lie in the crystallographic *bc*-plane generating a two-dimensional network. Propagation along the crystallographic *a* direction relies on a close N···H interaction at 3.7 Å to an aromatic-H atom, and the degree of spin delocalisation onto the C_6 ring [51].

4.1.1.3
Magnetic Behaviour

The room temperature moment of **4b** (like the dithiadiazolyl derivatives, **2**, described above) is 1.3 μ_B close to, but a little less than that expected for an

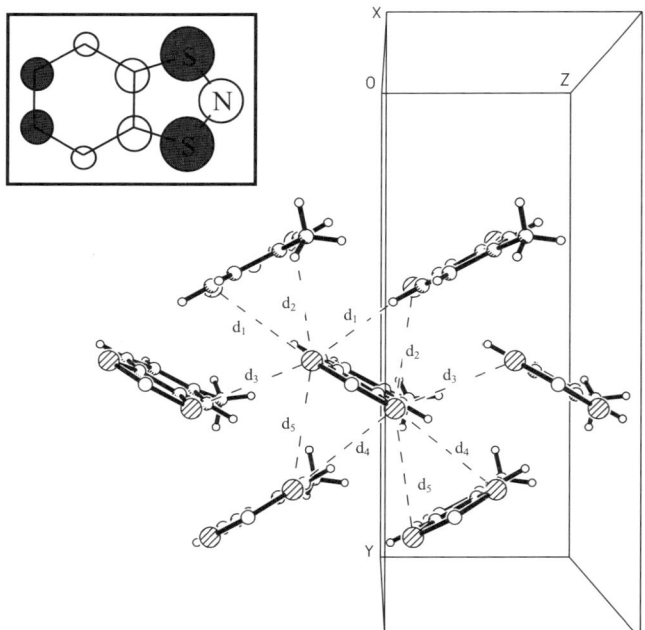

Fig. 13. Structure of **4b** with (inset) the symmetry of the SOMO of the closely related radical, **4a**

$S = 1/2$ paramagnet. This is entirely consistent with the presence of strong antiferromagnetic exchange interactions between radical centres, a common theme of thiazyl radical chemistry. The susceptibility passes through a broad maximum at 140 K indicative of the onset of low-dimensional antiferromagnetic order. The exceptionally high temperature for χ_{max} is indicative of the strength of the antiferromagnetic exchange interactions between radicals. Below 25 K the susceptibility curve increases rather sharply most likely due to the contribution from a small mole fraction of non correlated paramagnetic centres in the sample (<1% radical defects). The low dimensionality of the high temperature magnetic behaviour is consistent with the two-dimensional nature of the thiazyl interactions observed in the solid state structure. An analysis of the high temperature susceptibility as a two-dimensional square-planar Heisenberg model of $S = 1/2$ centres provided a good agreement between calculated and experimental data with $J = -72$ K. The adequacy of the fit is reflected in both the position and height of the maximum in χ. In Fig. 14 the magnetic susceptibility of **4b**, corrected for diamagnetic and uncorrelated paramagnetic contributions is represented in reduced units and compared with representative examples of the $S = 1/2$ square planar Heisenberg model as taken from the literature [51].

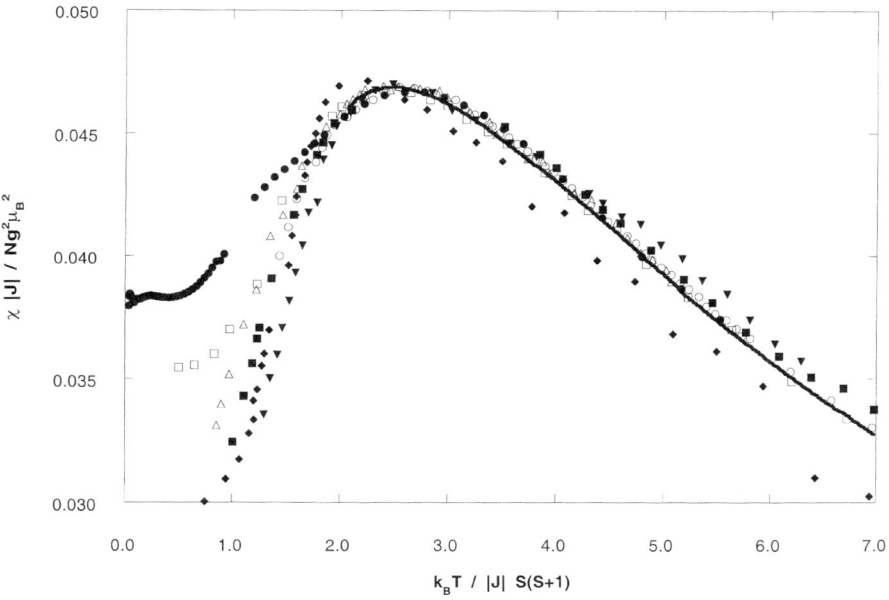

Fig. 14. Experimental temperature dependence of the susceptibility, in reduced units, of **4b** (●) compared to representative examples of the $S = 1/2$ square planar Heisenberg model; [Cu(C$_5$H$_5$NO)$_6$(BF$_4$)$_2$] (△), CuF$_2 \cdot$ 2H$_2$O (□), (2-NH$_2$-5-Cl-C$_5$H$_5$N)$_2$CuBr$_4$ (○), (2-NH$_2$-5-CH$_3$-C$_5$H$_5$N)$_2$CuBr$_4$ (■), [Cu(pz)$_2$-(NO$_3$)][PF$_6$] (◆), [Cu(pz)$_2$][ClO$_4$] (▼). Data are scaled to the high temperature series prediction (solid line)

4.1.2
Naphthalene-1,3,2-Dithiazolyl, NDTA

4.1.2.1
Crystal Structure

The crystal structure of NDTA, **4c**, [triclinic, *P-1*] was reported [52] by Oakley and contains two molecules in the asymmetric unit. The molecules are packed in a close-packed herringbone arrangement similar to that observed for **4b** and the closest intermolecular contacts are in the range 3.60–3.87 Å, a little shorter than those observed in **4b**.

4.1.2.2
Electronic and Magnetic Behaviour

The EPR spectra of **4c** (g = 2.0067, a_N = 11.06G) indicates significant localisation of the unpaired spin density on the heterocyclic ring [53]. An analysis of the susceptibility of **4c** indicates that at room temperature it is paramagnetic with a magnetic moment of ca. 1.5 μ_B analogous to **4b**. On cooling the susceptibility passes [52] through a broad maximum around 200 K before undergoing a phase transition to a more strongly antiferromagnetically coupled state. It passes through another maximum in its susceptibility around 25 K with eventual loss of paramagnetism [53].

4.2
1,3,2-Dithiazolyl Radicals with π-Stacking

Considerable attention has been paid in recent years to the development of neutral radicals, such as **2** and **4**, in the design of conducting materials [54]. Overlap of the π^* SOMOs of these neutral radicals along the stacking direction has been proposed to lead to a conduction band [55]. The extent of overlap evidently depends on the closeness of the intermolecular separation along the stacking direction. If the overlap is small then the unpaired spins can be considered localised, whereas if the overlap is good a band forms and the magnetic properties more closely match those of a low-dimensional conductor [56]. In many instances a charge-density wave driven (CDW), Peierls distortion of the lattice occurs and the stack becomes irregular with alternating long and short contacts. This leads to a band gap and the material becomes insulating. From a magnetic viewpoint, this can be envisaged as a simple spin-paired dimerisation process and the resultant species is diamagnetic. This Peierls distortion is a phenomenological problem for many low-dimensional systems and a number of methods have been employed to inhibit this process (metal-insulator transition) [54]. Doping of the structure with halogens, for example, removes some electrons from the conduction band and this inhibits dimerisation, although more complex CDW-driven periodic distortions of the lattice can also occur to open up a band gap at the Fermi level [57]. Alternatively, maximising inter-stack interactions [39] can also prove to be beneficial since these provide an energy barrier to inhibit the distortion.

As a consequence, the magnetic properties of π-stacked radicals can often be complex, with magnetic moments significantly below that anticipated for $S = 1/2$ spins, the exact value depending on the extent of the development of the band structure.

4c   **4d**   **4e**

4f   **4g**

4.2.1
Quinoxaline-1,3,2-Dithiazolyl, QDTA

This radical, **4d**, was first reported by Wolmershauser in 1990 [58], a detailed EPR investigation of this radical was then reported in 1992 [59], but it was not until 1997 that its crystal structure was reported [52].

4.2.1.1
Electronic Structure

EPR studies of **4d** have, in contrast to other dithiazolyl radicals described above, shown considerable hyperfine coupling to substituents on the dithiazolyl ring; not only can N-coupling to the quinoxaline substituent be resolved, but also coupling to the four H atoms of the quinoxaline ring [59]. The second derivative EPR spectrum of **4d** at 222 K is shown in Fig. 15 and, unlike the majority of thiazyl radicals of this class, shows considerable fine structure due to hyperfine interactions with the benzo-fused ring.

4.2.1.2
Crystal Structure

The crystal structure of **4d** (monoclinic, $P2_1$) [52] exhibits a slipped π-stack motif (Fig. 15b). The molecules π-stack along the crystallographic a-axis with a separation (at 293 K) corresponding to the length of the crystallographic a-axis, 3.7105(8) Å. In addition to this π-π interaction, there is only one other contact within 4.0 Å: an interstack S···S contact at 3.84 Å.

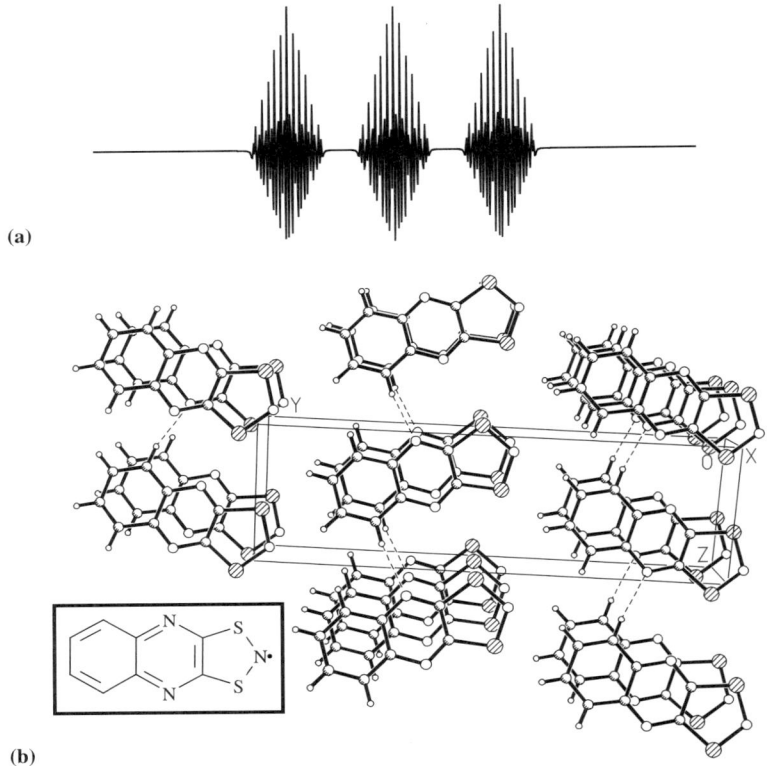

Fig. 15. (a) EPR spectrum of **4d** illustrating hyperfine interactions to the N and H atoms of the fused ring and (b) its solid state structure

4.2.1.3
Magnetic Properties of QDTA

The regular π-stacked structure gives rise to some degree of band formation, despite the relatively long intra-stack separation which is beginning to approach the sum of the van der Waals radii of S perpendicular to the ring plane, 4.06 Å [9]. At room temperature the magnetic moment corresponds to just 0.9 μ_B or 30% unpaired spins. On cooling the number of unpaired spins decreases steadily such that at 120 K all paramagnetism is quenched [53].

4.2.2
The Benzo-1,2:4,5-Bis(Dithiazolyl) Biradical, BBDTA

This compound, **4e**, was initially reported independently by Wudl [60] and Wolmershauser in 1986 [61]. Further work by Wudl provided a detailed EPR study [62]. This was followed in 1998 by further work by Wolmershauser who had partially oxidised the diradical to the radical cation and had isolated the tetrachloroferrate salt [63]. More recently, Oakley and co-workers have published the structure of **4e** and provided a detailed characterisation of the electronic structure [64].

4.2.2.1
Electronic Structure

The electronic structure of **4e** has been the subject of some considerable debate. Theoretical calculations on the benzo-fused derivative BDTA indicate strong mixing of the b_1 SOMO of the dithiazolyl ring with the π-system of benzene [64]. This mixing also occurs in **4e**, although the extent and nature of the intramolecular exchange between radicals becomes important. Theoretical calculations (cep-31 + g** basis set) indicate that the singlet state lies slightly higher (175 cm^{-1}) in energy than the triplet [64]. The EPR spectrum of **4e** is both sample and solvent dependent [62, 64] with the signal intensity being very weak in relation to the sample concentration, suggesting the major species in solution is closed shell. Recent analysis by Oakley [64] suggests that considerable association of molecules occurs in solution, leading to either closed-shell dimers, or open-shell dimers in which the radical centres are well separated leading to a pair of equivalent, non-interacting doublets. Dissolution in toluene disfavours the association and a spectrum more closely resembling a pentet structure (superimposed with the triplet spectrum arising from the non-interacting spin doublets) is observed. The pentet structure is consistent with a diradical in which the intramolecular exchange coupling (J_{ex}) between radical centres is greater than the hyperfine coupling, a_N.

4.2.2.2
Crystal Structure

The crystal structure of **4e** is shown in Fig. 16. Diradical **4e** crystallises in the monoclinic space group $P2_1/c$ [64]. The structure consists of discrete molecules of **4e** which form strongly slipped stacks (i.e., the ring plane is not orthogonal to the stacking direction) along the crystallographic a-axis. The molecular plane is tilted at 32.4° to this stacking direction. The mean interplanar separation along the stacking direction is 3.49 Å, shorter than that observed for **4d**. Additional inter-stack interactions between heterocyclic rings range from 3.12–3.74 Å.

4.2.2.3
Electronic Properties

The compound is essentially diamagnetic at room temperature, although a diamagnetic correction reveals some unpaired spin density, equivalent to ca. 1% of Curie spins [64]. This is consistent with the much closer intra-stack separation 3.49 Å (cf. 3.71 Å for **4d**). Although there is a considerable tilt of the molecular plane to the stacking axis, the better overlap evidently gives rise to a more well-developed band structure. An analysis of the conductivity indicates that it is a semiconductor with a band gap of 0.22 eV [64].

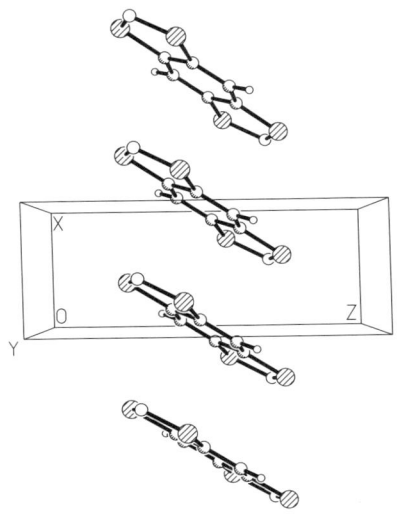

Fig. 16. Crystal structure of **4e**, viewed down the crystallographic *b*-axis

4.2.3
1,2,5-Thiadiazolo-Dithiazolyl-Pyrazine, TDP-DTA

4.2.3.1
Electronic Structure

The EPR spectrum of **4f** comprises a triplet of pentets due to coupling to the dithiazolyl N and the two pyrazine N atoms (g = 2.0071, a_N = 9.50G, a_N = 2.07G) [53]. The assignment of the smaller hyperfine coupling was made on the basis of spin density calculations at the MNDO level. These calculations indicate significant delocalisation of the SOMO onto the pyrazine substituent and considerably more than that observed for the benzo-derivative, **4a** [53].

4.2.3.2
Crystal Structure

At room temperature **4f** crystallises in the triclinic space group P-1 with two molecules in the unit cell [53]. The radicals form a slipped π-stack structure along the crystallographic c axis, but with significant slippage, such that equivalent atoms are separated by some 4.45 Å, considerably beyond the sum of the van der Waals radii of S.

On cooling to 150 K, a phase transition is observed. The low temperature phase has the same space group but doubles in size, with now four molecules in the unit cell [53]. In the low temperature phase the structure is built up of dimers which are linked through three S···S contacts in the range 3.40–3.48 Å. The change in structure upon cooling has profound effects on the magnetic behaviour. The phase change is reversible and on warming the high temperature phase is reformed. However, the temperature at which the phase change occurs on cooling (broad transition between 150 K and 50 K, based on magnetic response) is different from the transition on warming (abrupt transition at 175 K). The compound thus offers a region of bistability between 50 and 150 K in which the magnetic behaviour is temperature dependent. The crystal structures of both high and low temperature phases are shown in Fig. 17.

4.2.3.3
Magnetic Properties of TDP-DTA

The room temperature moment for **4f** is ca. 1.3 μ_B, a little lower than that expected for an $S = 1/2$ paramagnet, but consistent with previous results on thiazyl radicals in which there is strong antiferromagnetic exchange [53]. On cooling below room temperature, the susceptibility passes through a broad maximum around 200 K consistent with the onset of short range antiferromagnetic order. Around 150 K, χ begins to drop markedly reaching a value of ~400 emu/mol at 50 K, equivalent to ca. 10% unpaired spins. Although the structural information clearly indicates a dimerisation process with quenching of the unpaired spins, the compound never becomes diamagnetic. Instead there is essentially a constant 10% of unpaired spins below 50 K which follow Curie–Weiss behaviour with a Weiss constant of −34 K. Oakley and co-workers indicate an antiferromagnetic state, although T_N is unspecified [53]. Evidently further studies are warranted to examine the detailed magnetic behaviour of this radical.

On warming the susceptibility remains essentially constant up to 170 K, but then begins to rise steeply. At 200 K the magnetic data coincides with the high temperature measurements once again [53].

This is, to our knowledge, the first reported observation of bistability in a thiazyl radical. The variation in susceptibility between the two phases is

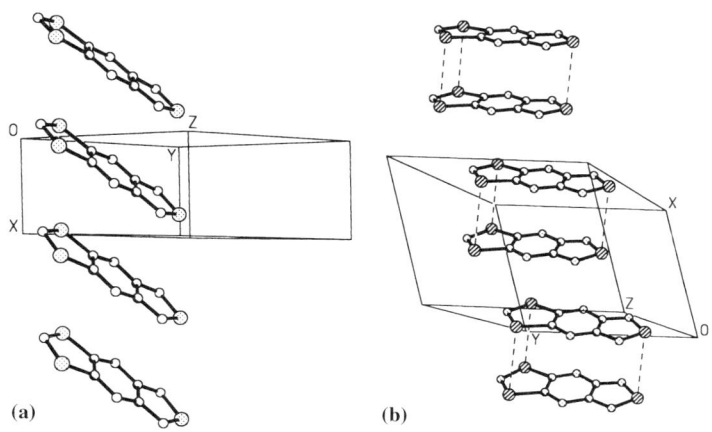

Fig. 17. High temperature (a) and low temperature (b) phases of **4f**

most marked around 150 K at which point the susceptibility for the high temperature phase is approximately twice that of the low temperature phase.

4.2.4
Trithiatriazapentalenyl Radical, TTTA

The radical **4g** was first reported in 1989 by Wolmershauser [65]. Little had been reported since, apart from an EPR study [66], until 1999 when Fujita and Awaga reported that this radical exhibited bistability at room temperature [67]. Other studies have since been reported [68].

4.2.4.1
Electronic Structure

EPR studies by us [68] and other workers [66], coupled with DFT calculations [68], have provided a clear understanding of the electronic structure of **4g**. The resolution of hyperfine coupling to both types of N atom in the dithiazolyl and thiadiazolyl ring has permitted an accurate analysis of the unpaired spin density [68]. The anisotropy in the g-values coupled with large hyperfine coupling to the high field g_1 component in the frozen glass spectra clearly indicate a π-based radical. An analysis of the spin density distribution provides a good correlation with theoretical calculations. These indicate that the majority of the unpaired spin density resides on the dithiazolyl ring, with approximately 22% of the unpaired spin density localised on the NSN unit (Fig. 18).

4.2.4.2
Crystal Structure

This radical exhibits bistability and its structure has been investigated by several groups. The structure of the high temperature phase was originally reported by Wolmershauser [65]. Its structure at room temperature was reported by Fujita [67] and variable temperature studies down to the phase transition temperature at 225 K have been made by Rawson and co-workers [68]. The low temperature phase has been reported at room temperature by Fujita and at 150 K by Rawson [68].

The high temperature phase (monoclinic $P2_1/c$) exhibits a regular π-stacking arrangement along the crystallographic *b*-axis. The intermolecular separation coincides with the crystallographic *b*-axis (3.71 Å) and the radicals

Fig. 18. Singly-occupied molecular orbital of **4g**

are inclined to the stacking direction by 21°. In addition to the vertical stacking, there are a number of lateral S···S and S···N interactions in the range 3.05–3.64 Å [68]. These are illustrated in Fig. 19.

In the low temperature phase (triclinic, P-1), the radicals dimerise with intermolecular S···S contacts in the range 3.24–3.32 Å, with slightly closer contacts between S atoms at the dithiazolyl ring which bears the larger spin density. The disruption of the regular π-stack motif is accompanied by a slipping of the dimer pairs away from a perfect stacking arrangement. A set of lateral interactions in the range 2.92–3.65 Å are observed. These are generally shorter than those observed in the high temperature phase [68]. The differences in the π-stacking arrangement of the high temperature and low temperature phases are shown in Fig. 20.

X-ray diffraction studies indicate that the high temperature phase undergoes a phase transition to the low temperature phase on cooling below 225 K. However, the reverse process does not occur until 317 K. Thus, at room temperature the compound can exist in either of two interconvertible forms, i.e., it is bistable.

4.2.4.3
Magnetic Properties

The magnetic properties of **4g** are remarkable. The room temperature behaviour of **4g** is dependent on sample history. If it has been annealed at 340 K, then its magnetic behaviour is that of the high temperature phase [67].

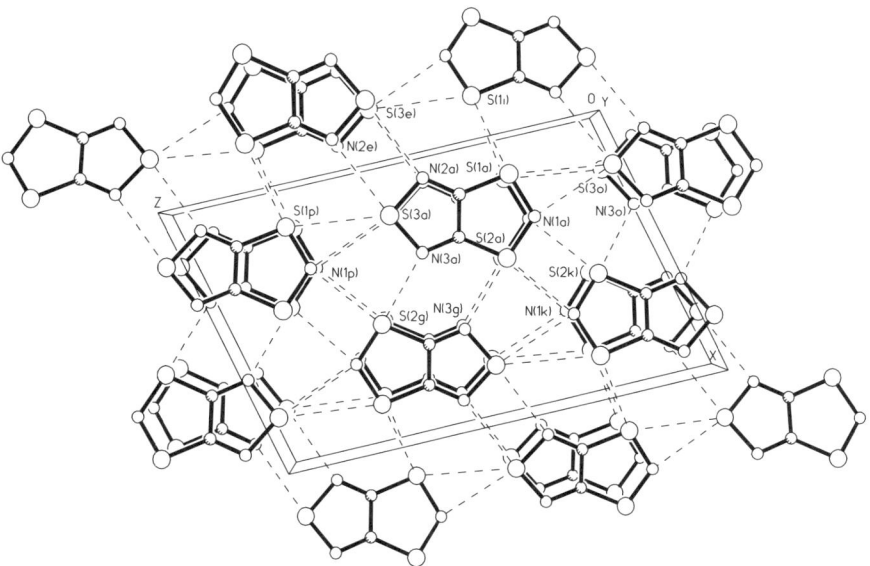

Fig. 19. Lateral contacts in the high temperature phase of **4g**, from ref. [68]

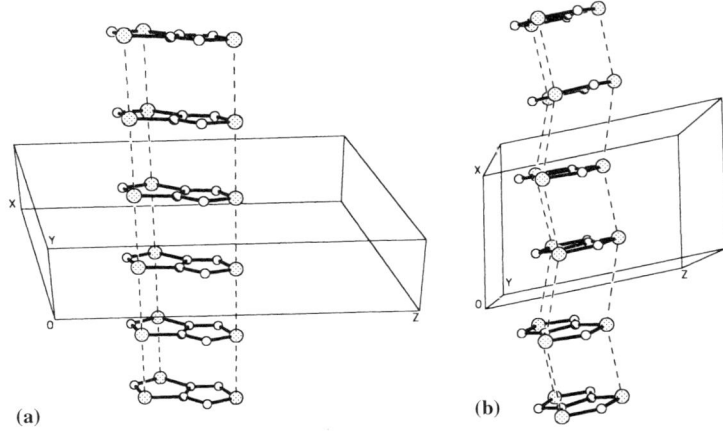

Fig. 20. π-stacking in (a) the high temperature and (b) low temperature phases of **4g**, from ref. [68]

Its room temperature moment is ca. $1.0\mu_B$, indicative of a very strong antiferromagnetically coupled regime. A fit of the data (350 to 230 K) to a one-dimensional antiferromagnetic chain yielded an intrachain exchange term, J_{intra}, of −320 K, but required a large inter-chain interaction, $J' = -60$ K [67]. The magnitude of J' to J_{intra} indicates that the one-dimensional model is not a good one, and the behaviour should be considered at least two-dimensionally. In conjunction with the low temperature crystallographic phase change, the susceptibility rapidly decreases below 230 K, and the compound becomes entirely diamagnetic by 150 K. The thermal hysteresis of **4g** (measured by EPR spectroscopy) is shown in Fig. 21.

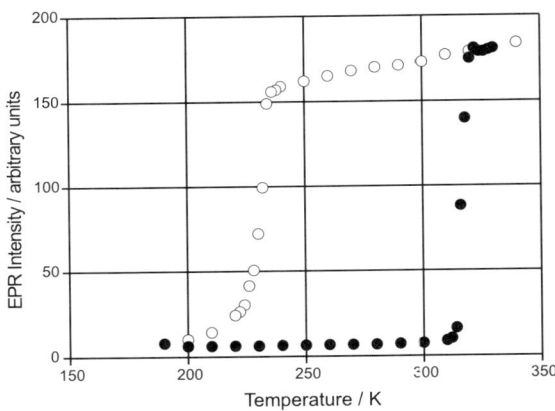

Fig. 21. Thermal hysteresis observed in **4f**, measured by EPR spectroscopy from ref. [68]

5
Comment

5.1
Is It Possible to Make Room Temperature Organic Magnets?

The magnetic behaviour of thiazyl radicals is characterised by strong antiferromagnetic exchange interactions between radical centres. The exchange interactions can be so strong that short-range order is often observed in the region 10^1–10^2 K. Using a mean field theory approximation, the preparation of organic radicals that order at room temperature seems an achievable goal. In the majority of cases it would appear likely that they will be antiferromagnets since they exhibit strong antiferromagnetic exchange interactions. However, the lower symmetry of many molecular structures may favour the formation of spin-canted antiferromagnets, as has already been observed for p-NCC$_6$F$_4$CNSSN, **2m**. In these circumstances a small spontaneous magnetisation may be observed.

However, for this goal to be achieved, the dimerisation energy associated with many of these radicals must be overcome and the radicals must pack in such a way that magnetic exchange interactions can propagate throughout the solid. The control of solid state structure is a complex and arduous task, but one which provides a goal worthy of achieving in the new millennium.

5.2
Bistability in Organic Radicals

Two dithiazolyl radicals (**4e** and **4f**) show thermal hysteresis; **4e** exhibits thermal hysteresis between a paramagnetic state and an antiferromagnetic phase [53], whereas **4f** exhibits thermal hysteresis between paramagnetic and diamagnetic states [67, 68]. A number of other dithiazolyl radicals have shown evidence that they undergo phase transitions, although the reversibility of their behaviour has not been reported.

The bistability in these systems presumably arises through competition between the spin-density wave driven dimerisation process and lattice packing forces. In the case of derivatives of **4**, the dimerisation energy is approximately zero and both open-shell (two doublet states) and the closed-shell configurations must have similar stabilities. The energy to interconvert between structures depends on the energy barrier for structural reorganisation. If the energy barrier is too high then two non-interconvertible polymorphs can be expected. Polymorphism has been observed in a number of thiazyl and selenazyl radicals, including **2f**. In these cases the crystal packing in the different phases is substantially different and interconversion is not possible in the solid state. For smaller energy barriers, i.e., requiring minimum lattice reorganisation, then interconversion can be expected. In the case of π-stacked radicals the reorganisation energy can be expected to be low since it requires only minor displacements along the stacking direction. Contraction or expansion of the crystal lattice on cooling or warming can lead to a variation in the relative stabilities of the two polymorphs.

In the case of the dithiadiazolyl radicals, 2, the enthalpy for dimerisation clearly favours dimeric structures over monomeric ones and (in the majority of instances) the dimeric phase is only observed; the asymmetry in the potential energy wells inhibits bistability. Interestingly, a number of regular π-stacked derivatives of 2 have recently been prepared [69] and it will be of interest to see whether, with appropriate control, bistability can be induced in dithiadiazolyl radicals.

The observation of bistability at room temperature in 4f is evidently of significance in the development of organic data recording devices. The development of new thiazyl radicals with different ranges of bistability will be of significant interest. Moreover, the methods which might be employed to switch the bistability in such devices will lead to exciting new avenues in materials chemistry.

6
References

1. Lahti PM (ed) (1999) Magnetic Properties of Organic Materials, Marcel-Dekker Inc, New York; Itoh K, Kinoshita M (eds) (2000) Molecular Magnetism: New Magnetic Materials, Gordon and Breach, Amsterdam
2. Turek P, Nozawa K, Shiomi D, Awaga K, Inabe T, Maruyama Y, Kinoshita M (1991) Chem Phys Lett 180: 327
3. Allemand PM, Khemani KC, Koch A, Wudl F, Holczer K, Donovan S, Grüner G, Thompson JD (1991) Science 253: 301
4. Nakatsuji S, Anzai H (1997) J Mater Chem 7: 2161
5. Demarcay E (1880) Compt Rend 91: 854
6. See, for example: (a) Banister AJ, Clarke HG, Rayment I, Shearer HMM (1974) Inorg Nucl Chem Lett 10: 647; (b) Gillespie RJ, Ireland PR, Vekris JE (1975) Can J Chem 14: 498; (c) Fairhurst SA, Johnson KM, Sutcliffe LH, Preston KF, Banister AJ, Hauptman ZV, Passmore J (1986) J Chem Soc Dalton Trans 1465
7. Krebs B, Henkel G, Pohl S, Roesky HW (1980) Chem Ber 113: 226; Gleiter R, Bartetzko R, Hoffmann P (1980) Z Naturforsch 35B: 1166; Gillespie RJ, Kent JP, Sawyer JF (1981) Inorg Chem 20: 3784; Thewalt U, Burger M (1981) Z Naturforsch 36B: 293; Small RWH, Banister AJ, Hauptman ZV (1984) J Chem Soc Dalton Trans 1377; Ayres B, Banister AJ, Coates PD, Hansford MI, Rawson JM, Rickard CEF, Hursthouse MB, Malik KMA, Motevalli M (1992) J Chem Soc Dalton Trans 3097
8. Emsley J (1995) The Elements, second edition, Oxford University Press
9. The van der Waals radii for S have been shown to be considerably aspherical with the van der Waals radius in the plane being ca. 1.60 Å, but ca. 2.03 Å perpendicular to the ring; see: Nyburg SC, Faerman CH (1985) Acta Crystallogr 41: 274
10. Selenium analogues of $S_3N_2^{+\bullet}$ can be found in the following references: (a) Awere EG, Brooks WVF, Passmore J, White PS, Sun X, Cameron TS (1993) J Chem Soc Dalton Trans 2439; (b) Awere EG, Passmore J, Preston KF, Sutcliffe LH (1988) Can J Chem 66: 1776
11. A number of C/N/Se radicals have also been prepared. The diselenadiazolyl radicals have been developed by Oakley and co-workers, see, for example, (1992) J Am Chem Soc 114: 1729 and references therein. Selenium analogues of the dithiazolyl radicals have also been reported, see Rawson JM, McManus GD (1999) Coord Chem Rev 189: 135 and references therein

12. (a) Genin H, Hoffmann R (1998) Macromol 31: 444; (b) Banister AJ, Gorrell IB, Clegg W, Jörgenson KA (1991) J Chem Soc Dalton Trans 1105
13. (a) Hofs HU, Bats JW, Gleiter R, Hartmann G, Mews R, Eckert-Maksic M, Oberhammer H, Sheldrick GM (1985) Chem Ber 118: 3781; (b) Boere RT, Oakley RT, Reed RW, Westwood NPC (1989) J Am Chem Soc 111: 1180; (c) Cordes AW, Goddard JD, Oakley RT, Westwood NPC (1989) J Am Chem Soc 111: 6147; (d) Cordes AW, Bryan CD, Davis WM, De Laat RH, Glarum SH, Goddard JD, Haddon RC, Hicks RG, Kennepohl DK, Oakley RT, Scott SR, Westwood NPC (1993) J Am Chem Soc 115: 7232; (e) Boeré RT, Moock KH (1995) J Am Chem Soc 117: 4755
14. Palacio F, Antorrena G, Castro M, Burriel R, Rawson JM, Smith JNB, Bricklebank N, Novoa J, Ritter C (1997) Phys Rev Lett 79: 2336
15. Rawson JM, Banister AJ, May I (1994) Magn Reson Chem 32: 487
16. See, for example: (a) Vegas A, Perez-Salazaar A, Banister AJ, Hey RG (1980) J Chem Soc Dalton Trans 12; (b) Cordes AW, Haddon RC, Hicks RG, Oakley RT, Palstra TTM (1992) Inorg Chem 31: 1802; (c) Boeré RT, Moock KH, Parvez M (1994) Z Anorg Allg Chem 620: 1589
17. Banister AJ, Hansford MI, Hauptman ZV, Wait ST, Clegg W (1989) J Chem Soc Dalton Trans 1705; (b) Bridsdon JN, Copp SB, Shriver MJ, Shuguang Z, Zawarotko M (1994) Can J Chem 72: 1143
18. Banister AJ, Batsanov AS, Dawe OG, Herbertson PL, Howard JAK, Lynn S, May I, Smith JNB, Rawson JM, Rogers TE, Tanner BK, Antorrena G, Palacio F (1997) J Chem Soc Dalton Trans 2539; Beer L, Cordes AW, Myles DJT, Oakley RT, Taylor NT (2000) Cryst Eng Comm 20
19. (a) Barclay TM, Cordes AW, George NA, Haddon RC, Itkis ME, Oakley RT (1999) Chem Commun 2269; (b) Bricklebank N, Hargreaves S, Spey SE (2000) Polyhedron 19: 1163
20. Banister AJ, Rawson JM, Clegg W, Birkby SL (1991) J Chem Soc Dalton Trans 1099
21. (a) Burford N, Passmore J, Schriver MJ (1986) J Chem Soc Chem Commun 140; (b) Brooks WVF, Burford N, Passmore J, Schriver MJ, Sutcliffe LH (1987) J Chem Soc Chem Commun 69; (c) Passmore J, Sun X (1996) Inorg Chem 35: 1313
22. Aherne CM, Banister AJ, Luke AW, Rawson JM, Whitehead RJ (1993) J Chem Soc Dalton Trans 1421
23. Passmore J, Sun X, Parsons S (1992) Can J Chem 70: 2972
24. (a) Wolmershauser G, Kraft G (1989) Chem Ber 122: 385; (b) Wolmershauser G, Kraft G (1990) Chem Ber 123: 881
25. Awere EG, Burford N, Haddon RC, Parsons S, Passmore J, Waszczak JV, White PS (1990) Inorg Chem 29: 4821
26. Barclay TM, Beer L, Cordes AW, Oakley RT, Preuss KE, Taylor NJ, Reed RW (1999) Chem Commun 531
27. Davies JE, Less RJ, May I, Rawson JM (1998) New J Chem 763
28. Awere EG, Burford N, Mailer C, Passmore J, Schriver MJ, White PS, Banister AJ, Oberhammer M, Sutcliffe LH (1987) J Chem Soc Chem Commun 66
29. Gleiter R, Bartetzko R, Hofmann P (1980) Z Naturforsch 35B: 1166
30. Yoshizawa K (1999) In: Magnetic Properties of Organic Materials, Lahti PM (ed), Marcel-Dekker Inc, New York, Chap 19
31. Beekman RA, Boeré RT, Moock KH, Parvez M (1998) Can J Chem 76: 85
32. Banister AJ, Lavender I, Rawson JM, Clegg W, Tanner BK, Whitehead RJ (1993) J Chem Soc Dalton Trans 1993: 1421
33. Bryan CD, Cordes AW, Fleming RM, George NA, Glarum SH, Haddon RC, MacKinnon CD, Oakley RT, Palstra TTM, Perel AS (1995) J Am Chem Soc 117: 6880
34. Alonso PJ, Antorrena G, Martinez JI, Novoa JJ, Palacio F, Rawson JM, Smith JNB (2000) J Chem Phys A (submitted)
35. Antorrena G, Davies JE, Hartley M, Palacio F, Rawson JM, Smith JNB, Steiner A (1999) Chem Commun 1393
36. Miller JS (1998) Adv Mater 10: 1553

37. Cordes AW, Haddon RC, Hicks RG, Oakley RT, Palstra TTM, Schneemeyer LF, Wasczak JV (1992) J Am Chem Soc 114: 1729
38. Banister AJ, Bricklebank N, Clegg W, Elsegood MRJ, Gregory CI, Lavender I, Rawson JM, Tanner BK (1995) J Chem Soc Chem Commun 679
39. Bryan CD, Cordes AW, Haddon RC, Hicks RG, Kennepohl DK, MacKinnon CD, Oakley RT, Palstra TTM, Perel AS, Scott SR, Schneemeyer LF, Wasczak JV (1994) J Am Chem Soc 116: 1205
40. Banister AJ, Bricklebank N, Lavender I, Rawson JM, Gregory CI, Tanner BK, Clegg W, Elsegood MRJ, Palacio F (1996) Angew Chem Int Ed Engl 35: 2533
41. Caro J, Fraxedas J, Santiso J, Figueras A, Rawson JM, Smith JNB, Antorrena G, Palacio F (1999) Thin Solid Films 352: 102
42. Rawson JM (2000) (unpublished results)
43. Langley PJ, Rawson JM, Smith JNB, Schuler M, Bachmann R, Schweiger A, Palacio F, Antorrena G, Gescheidt G, Quintel A, Rechsteiner P, Hulliger J (1999) J Mater Chem 9: 1431
44. Palacio F, Antorrena G, Castro M, Burriel R, Rawson JM, Smith JNB, Bricklebank N, Novoa J, Ritter C (1997) Phys Rev Lett 79: 2336
45. Carlin RL (1989) Magnetochemistry, Springer, Berlin
46. The estimated canting angle is based on the g-value from recent EPR studies (ref [34])
47. Goeta AE, Antorrena G, Palacio F, Rawson JM, Smith JNB (2001) (unpublished work)
48. Pratt FL, Goeta AE, Palacio F, Rawson JM, Smith JNB (2000) Physica B 289: 119
49. Less RJ, Rawson JM, Mito M, Takagi S, Deguchi H, Takeda K, Palacio F (2001) (manuscript in preparation).
50. Wolmershauser G, Schnauber M, Wilhelm T (1984) J Chem Soc Chem Commun 573
51. McManus GD, Rawson JM, Feeder N, Palacio F, Oliete P (2000) J Mater Chem 10: 2001
52. Barclay TM, Cordes AW, George N, Haddon RC, Oakley RT, Palstra TTM, Patenaude GW, Reed RW, Richardson JF, Zhang H (1997) Chem Commun 873
53. Barclay TM, Cordes AW, George NA, Haddon RC, Itkis ME, Mashuta MS, Oakley RT, Patenaude GW, Reed RW, Richardson JF, Zhang H (1998) J Am Chem Soc 120: 352
54. Cordes AW, Haddon RC, Oakley RT (1994) Adv Mater 6: 798
55. Haddon RC (1975) Nature 256: 394
56. Peierls RE (1953) Quantum Theory of Solids, Oxford, London
57. Bryan CD, Cordes AW, Fleming RM, George NA, Glarum SH, Haddon RC, MacKinnon CD, Oakley RT, Palstra TTM, Perel AS (1995) J Am Chem Soc 117: 6880
58. Wolmershauser G, Kraft G (1990) Chem Ber 123: 881
59. (a) Chung Y-L, Fairhurst SA, Gillies DG, Kraft G, Krebber AML, Preston KF, Sutcliffe LH, Wolmershauser G (1992) Magn Reson Chem 30: 774; (b) Chung Y-L, Fairhurst SA, Gillies DG, Preston KF, Sutcliffe LH (1992) 30: 666
60. Williams KA, Nowak MJ, Dormann E, Wudl F (1986) Synth Met 14: 233
61. Wolmershauser G, Schnauber M, Wilhelm T, Sutcliffe LH (1986) Synth Met 14: 239
62. Dormann E, Nowak MJ, Williams KA, Angus Jr. RO, Wudl F (1987) J Am Chem Soc 109: 2594
63. Wolmershauser G, Wortmann G, Schnauber M (1988) J Chem Res (S) 358
64. Barclay TM, Cordes AW, De Laat RH, Goddard JD, Jeter DY, Mawhinney RC, Oakley RT, Palstra TTM, Patenaude GW, Reed RW, Westwood NPC (1997) J Am Chem Soc 119: 2633
65. Wolmershauser G, Johann R (1989) Angew Chem Int Ed Engl 28: 920
66. Chung Y-L, Sandall JPB, Sutcliffe LH, Joly H, Preston KF, Johann R, Wolmershauser G (1991) Magn Reson Chem 29: 625
67. Fujita W, Awaga K (1999) Science 286: 261
68. McManus GD, Rawson JM, Feeder N, van Duijn J, McInnes EJ, Novoa J, Palacio F, Oliete P (2001) J Mater Chem 11 (in press)
69. Bricklebank N, Personal communication; Hargreaves S (2000) PhD Thesis, Sheffield Hallam Univ

Magnetism in Fullerene Derivatives

Denis Arčon[1], Kosmas Prassides[2]

[1] Institute Josef Stefan, Jamova 39, 1000 Ljubljana, Slovenia
  E-mail: denis.arcon@ijs.si
[2] School of Chemistry, Physics and Environmental Sciences, University of Sussex, Brighton BN1 9QJ, UK
  E-mail: k.prassides@susx.ac.uk

The discovery of C_{60} and other members of the fullerene family opened new horizons in the design and synthesis of novel materials. Particularly exciting have been the results of the synthesis of C_{60}-based materials, which exhibit transitions to ferro- and antiferro-magnetic ground states. In this chapter, we describe structural, electronic, and magnetic properties of fullerene salts in which magnetic transitions have been reported, i.e., the TDAE-C_{60} ferromagnet, the $(NH_3)K_3C_{60}$ antiferromagnet, and the AC_{60} and Na_2AC_{60} (A = K, Rb, Cs) polymers. The effects of different thermal treatments, changes of the interfullerene distance, and the role of orientational order/disorder of C_{60}^{n-} ions on the magnetic properties are discussed.

Keywords: Fullerenes, Ferromagnetism, Antiferromagnetism, Polymers, Metals, Metal-insulator transition, Magnetic resonance, Muon spin relaxation, Structural properties

1	Introduction	130
2	Ferromagnetism in TDAE-C_{60}	133
3	Antiferromagnetic Ordering in $(NH_3)K_3C_{60}$	140
4	Magnetism in Polymeric Fullerides	147
4.1	Magnetic Properties of the AC_{60} Family	147
4.2	Magnetic Properties of Na_2RbC_{60} and Related Salts	152
5	Summary and Future Prospects	159
6	References	160

List of Abbreviations

AFM	antiferromagnetic
AFMR	antiferromagnetic resonance
CDW	charge density wave
HF	high field
HOMO	highest occupied molecular orbital
LF	longitudinal field
LRO	long range order

LUMO   lowest unoccupied molecular orbital
μSR   muon spin relaxation
SC   superconducting
SDW   spin density wave
ZF   zero field

1
Introduction

Organic materials are molecular solids in which intermolecular distances are commonly much longer than the interatomic distances found in inorganic metallic and ferromagnetic solids. The molecules are normally diamagnetic, closed-shell species and as such they do not usually display interesting electronic and magnetic properties. Fortunately, in the past decade, there has been enormous progress in the development of molecular-based magnetic materials [1] with large molecular units as building blocks. In these, the magnetic moments are derived exclusively from unpaired p-electrons, as opposed to the more commonly encountered d- and f-electron systems. Basically, one can divide organic molecules which are open-shell entities into two groups containing either neutral, localised unpaired electron-containing functional groups (free electrons) or charged species (radical ions). The nitroxyls are examples of members of the former category. Radical ions result by electron transfer from the π-orbitals of an electron-rich molecule or transfer of an electron to the π molecular orbital of an electron-deficient molecule. In this chapter, we review new magnetic materials of the donor-acceptor type, based on the buckminsterfullerene, C_{60}, molecule.

A new chapter in organic chemistry and condensed matter physics was opened recently with the discovery of C_{60} (Fig. 1a) and other fullerenes [2]. In C_{60}, 60 carbon atoms are arranged in the form of a truncated icosahedron (soccer ball). This molecule has the high symmetry of an icosahedron (point group I_h) with six five-fold, ten three-fold axes, fifteen two-fold symmetry axes, and inversion symmetry. From an organic chemist's perspective, C_{60} represents an entirely new type of organic molecule and opens new horizons in the design and synthesis of novel compounds. On the other hand to a physicist,

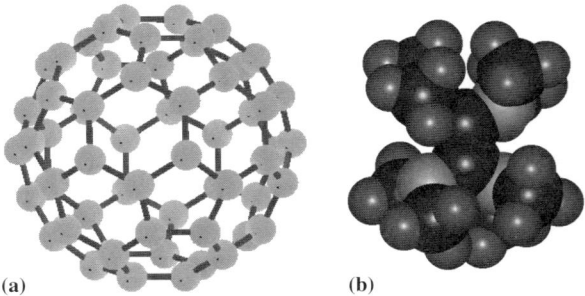

Fig. 1. Molecular structures of (a) C_{60} and (b) tetrakis-dimethyl-aminoethylene (TDAE)

the C_{60} molecule represents a highly symmetric, stable molecular analogue of a very large atom, which in the solid state can adopt a crystal structure with relatively large interstitial sites that can be filled with atomic or molecular host units (Fig. 2).

Even before its production in macroscopic quantities, the electronic structure of molecular C_{60} had attracted considerable theoretical interest. Due to the small overlap between the electron clouds of neighbouring molecules, the electronic structure of solid C_{60} can be described as a relatively small perturbation of its molecular orbital levels scheme [3]. Pristine solid C_{60} is an insulator and there is a ≈ 1.5 eV gap between the (filled) highest occupied molecular orbital (HOMO) and the lowest unoccupied molecular orbital (LUMO) (Fig. 3) [4]. The LUMO of t_{1u} symmetry is triply degenerate and can accept up to six electrons per C_{60} molecule. Intercalated C_{60} salts should thus become metallic if fewer than 6 electrons are injected into the t_{1u}-derived conduction band. This idea stimulated a great amount of work on intercalated C_{60} compounds, resulting in some spectacular results. Certainly one of the most important achievements has been the discovery of superconductivity with relatively high T_C in alkali fullerides with stoichiometry A_3C_{60}, where (A = Na, K, Rb, Cs) [5].

The importance of C_{60} as a new building block for the design of novel magnetic materials lies with its great flexibility:

(a) there is vast number of ways of doping pristine C_{60} that can result in materials with exciting electronic and magnetic properties;

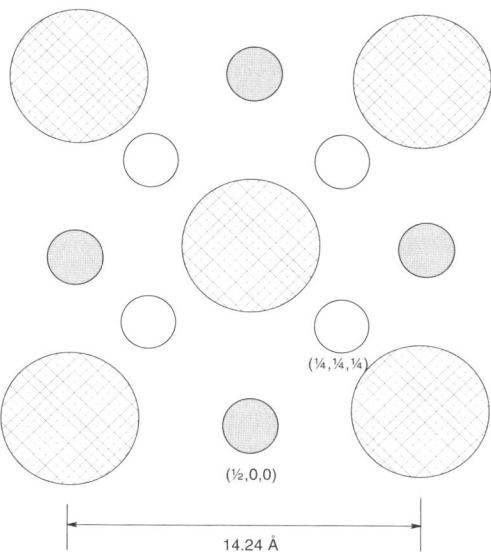

Fig. 2. The cryolite structural type adopted by the K_3C_{60} fulleride. The shaded and white spheres represent potassium ions residing in the octahedral and tetrahedral sites, respectively

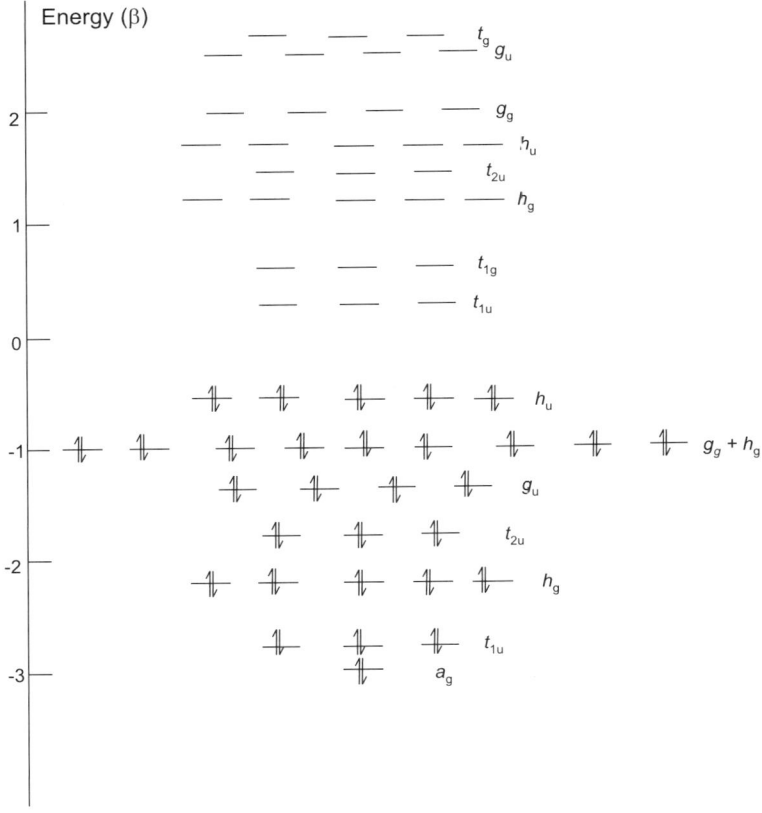

Fig. 3. Molecular orbital level scheme of C_{60} (adapted from ref. [4])

(b) the C_{60}–C_{60} distances can be easily controlled through the use of appropriate ions with different sizes (changing K^+ to Cs^+, for instance), of co-ordination complexes (co-intercalation of alkali ions and neutral NH_3) and of large organic donors;
(c) different thermal treatments of the samples can lead to stabilisation of competing phases characterised by the formation of interfullerene bridging C—C bonds; and
(d) the electronic and magnetic properties of the materials are sensitive to the presence of orientational order/disorder of C_{60} units.

Most attempts to synthesise intercalation compounds of C_{60} have employed the introduction of dopants into the empty interstitial space between the C_{60} units (Fig. 2). In some cases – the building blocks are shown in Fig. 1 – such

---

The ionic radii of the alkali ions are: 0.69 Å (Li^+), 0.95 Å (Na^+), 1.33 Å (K^+), 1.48 Å (Rb^+), and 1.69 Å (Cs^+). These should be compared with the size of the tetrahedral (1.12 Å) and octahedral (2.06 Å) interstices.

doping led to very interesting magnetic behaviour. In this review, we focus our attention only to those fulleride systems in which magnetic transitions have been encountered, i.e., the TDAE-C_{60} ferromagnet, the $(NH_3)K_3C_{60}$ antiferromagnet, and the AC_{60} and Na_2AC_{60} (A = K, Rb, Cs) polymers.

2
Ferromagnetism in TDAE-C_{60}

The purpose of this section is to review the magnetic properties of one of the most interesting new magnetic organic compounds of the past decade, namely the ferromagnetic fulleride tetrakis-dimethylamino-ethylene-C_{60} (TDAE-C_{60}) (Fig. 4), first discovered in 1991 by Wudl and co-workers at the University of California, Santa Barbara [6]. At the time of discovery, the Curie temperature, T_C = 16 K of TDAE-C_{60} was one order of magnitude higher than the existing record [7] and brought the research field of p-electron ferromagnetism from the realms of the esoteric into the mainstream.

When the Santa Barbara group measured the magnetic properties of TDAE-C_{60} in powder form, a steep rise in the real part of the ac susceptibility (Fig. 5) was evident below 16 K, implying the occurrence of a ferromagnetic-type transition at a surprisingly high transition temperature, T_C = 16 K for a compound comprising only elements of the first row (C, N, H) of the periodic table. Since then, numerous experiments have been performed on both powders and single crystalline samples of TDAE-C_{60}, leading to a comprehensive understanding of its intriguing behaviour.

It was soon realised that the ferromagnetic transition in TDAE-C_{60} was not of a conventional type, which could be simply described within some standard theory. Some of the puzzling properties of TDAE-C_{60} were described in detail in the original report on its synthesis and characterisation [6]:

(i) despite the sharp increase of the magnetisation below T_C, its temperature dependence did not follow conventional mean field theory behaviour;
(ii) within experimental error, no hysteresis was observed between measurements under cooling and heating protocols;
(iii) similarly, the coercive field and the remnant magnetisation were found to be zero; and
(iv) the spontaneous magnetisation was very small, leading to a low-temperature saturation moment of ~0.11 μ_B/C_{60}.

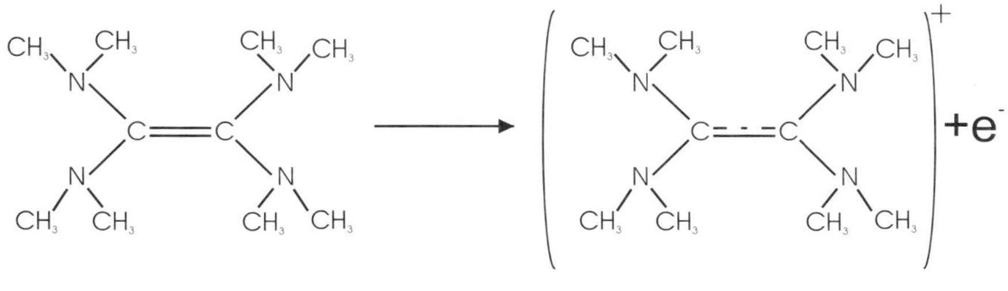

Fig. 4. Schematic diagram of the TDAE reduction

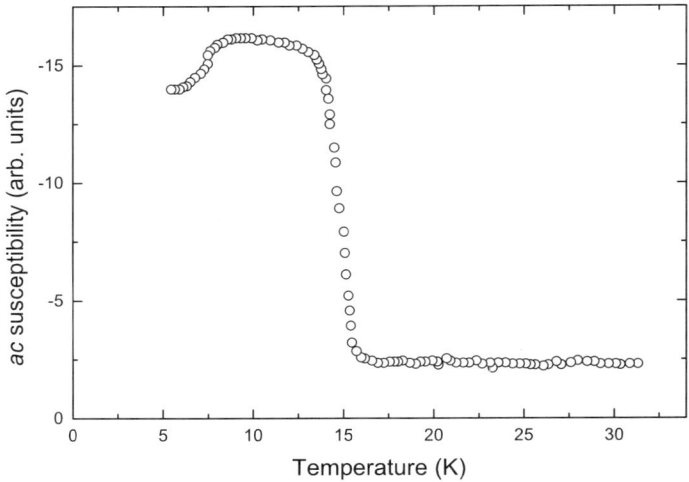

Fig. 5. Temperature dependence of the real part of the ac susceptibility (v = 3 kHz) of a single crystal of TDAE-C_{60}, showing the onset of the ferromagnetic transition at T_C = 16 K

In order to explain these unusual magnetic properties, a variety of models appeared in the early literature, describing TDAE-C_{60} as an itinerant ferromagnet [6], a superparamagnet [8], a spin-glass [9], or a weak ferromagnet [10].

When TDAE (a liquid at room temperature) is added to C_{60} in solution, the resulting salt crystallises as small particles, which can be removed by filtration. $TDAE^+C_{60}^-$ is a charge-transfer salt with a monoclinic structure, different from the high-symmetry cubic structures of pristine C_{60} and its alkali intercalated derivatives [11]. The crystal structure of TDAE-C_{60} was first determined by Stephens et al. [12] to be monoclinic (space group $C2/m$) with two formula units per unit cell. However, a subsequent X-ray single-crystal structural determination [10] identified the existence of a superlattice, arising from the presence of symmetry-inequivalent $TDAE^+$ units in the unit cell. Two subcells are present, with the $TDAE^+$ ions shifting along the b axis by ~0.02 Å in opposite directions and resulting in doubling of the unit cell along the c axis (Fig. 6). The structure at room temperature is then monoclinic (space group $C2/c$) with cell dimensions a = 15.858(2) Å, b = 12.998(2) Å, c = 19.987(2) Å, β = 93.37(3)° and four formula units per unit cell. The TDAE units reside in the (1/2, 1/2 + δ, 3/4), (1/2, 1/2−δ, 1/4), (0, δ, 3/4), and (0, −δ, 1/4) sites with δ = 0.002 and the C=C double bond parallel to the c axis, whereas the C_{60} molecules are centred at the (0, 1/2, 0), (0, 1/2, 1/2), (1/2, 0, 0), and (1/2, 0, 1/2) sites. The resulting C_{60}-C_{60} centre-to-centre distances are shortest along the crystal c axis (9.99 Å at room temperature as compared to 10.24 Å in solid C_{60} [11]). At 80 K, the centre-to-centre distances along the c-axis contract further to 9.87 Å. The unusually short distances between adjacent C_{60} units are crucial in leading to coupling between spins localised on neighbouring C_{60}^- ions and the formation of the ferromagnetically correlated spin state.

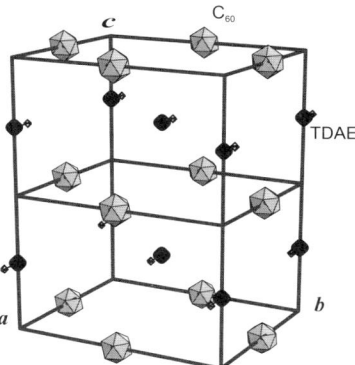

Fig. 6. Schematic diagram of the monoclinic unit cell of TDAE-C_{60} (space group $C2/c$) consisting of two subcells stacked along the c axis. The arrows indicate the small shifts of the TDAE units along the b direction

Since charge transfer from TDAE to C_{60} leads to a partially full molecular orbital (Fig. 3) and because of the close contacts of the C_{60} units, it might be expected that TDAE-C_{60} is a metal with a highly anisotropic electronic structure, displaying itinerant ferromagnetism [6]. However, powder infrared absorption spectroscopy revealed that TDAE-C_{60} displays no Drude tail and exhibits no absorption at low frequencies [13], implying that the material is most probably an insulator. Microwave absorption measurements [14] also supported the insulating character of TDAE-C_{60}. The issue was finally settled unambiguously by dc and ac conductivity measurements on single crystals of TDAE-C_{60} as a function of temperature [15]. These experiments ruled out the possibility of an itinerant ferromagnetic ground state below $T_C = 16$ K.

The first unambiguous evidence for long-range magnetic order came from zero-field muon spin relaxation (ZF-μ^+SR) experiments performed on TDAE-C_{60} powder [16]. μ^+SR spectroscopy is an extremely powerful technique in cases of small-moment magnetism and in all instances where magnetic order is of a random, very short-range, spatially inhomogeneous or incommensurate nature [17]. The μ^+ are implanted into the solid sample and after they come to rest at an interstitial site, they act as highly sensitive microscopic local magnetic probes. In the presence of local magnetic fields, $\langle B_\mu \rangle$, they will precess with a frequency $v_\mu = (\gamma_\mu/2\pi)\langle B_\mu \rangle$, where $\gamma_\mu/2\pi = 13.55$ kHz/G. In the absence of an applied external field (ZF), the appearance of a precession signals the onset of an ordering (ferromagnetic or antiferromagnetic) transition. Moreover, application of a magnetic field parallel to the initial muon spin polarisation (LF) allows the decoupling of the μ^+ spin from the static internal fields.

In well-annealed powder TDAE-C_{60}, a heavily damped oscillating signal [$v_\mu = 0.92(2)$ MHz at 3.2 K] was observed in zero field (Fig. 7), providing definite proof of the existence of long-range magnetic order below 16.1 K. The strong muon spin relaxation – whose quasi-static nature was confirmed by LF

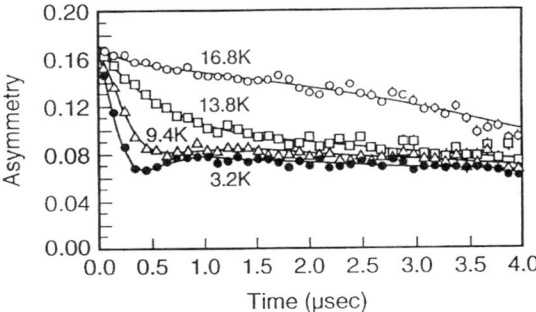

Fig. 7. Evolution of the zero-field (ZF) μ^+ spin polarisation, $P_\mu(t)$ between 3.2 and 16.8 K for TDAE-C_{60}

measurements – found in ZF implies the presence of substantial spatial disorder and inhomogeneity effects. The local field distribution, $\langle \Delta B^2 \rangle^{1/2} = 48(2)$ G is only smaller than $\langle B_\mu \rangle = 68(1)$ G by a factor of 1.4. In addition, the temperature dependence of v_μ (Fig. 8) which also mirrors that of the magnetisation measured at 100 G behaves on approaching T_C in a different manner from that expected by the mean-field treatment of a 3D Heisenberg exchange model. It follows well Bloch's $T^{3/2}$-law and thus it appears that spin wave (magnon) excitations dominate the magnetic behaviour, even at temperatures close to T_C. Alternatively, the magnon-like behaviour may reflect a more complicated physical picture of the system, where the intrinsic orientational disorder associated with the fullerene molecules gives rise to a broad distribution of exchange constants (and ordering temperatures) [17, 18].

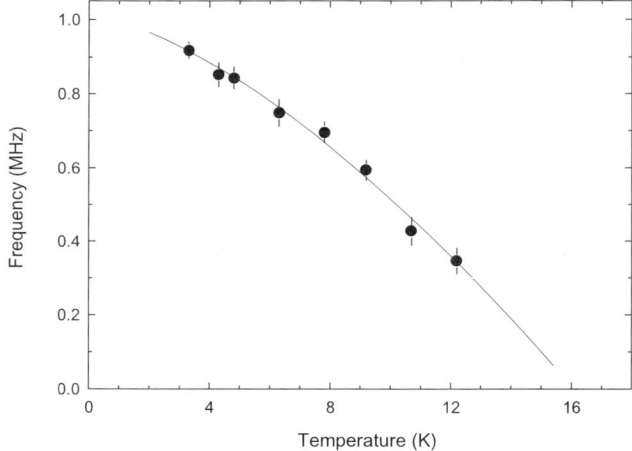

Fig. 8. Temperature dependence of the μ^+ precession frequency, v_μ (solid circles) in TDAE-C_{60} powder. The solid line is a fit to Bloch's $T^{3/2}$-law

μ^+SR thus provided clear evidence for the long-range ordering below $T_C = 16$ K, but many details were still missing. Additional information has been obtained from electron spin resonance (ESR) experiments. The ESR line width (ΔH_{pp}) in the paramagnetic phase gives information about the spin dynamics in the system under investigation. For TDAE-C_{60}, a dramatic decrease in ΔH_{pp} was detected below $T_O = 160$ K (Fig. 9). Above T_O, the line width is determined by the dipolar interactions between the neighbouring spins, while below T_O the exchange interactions become important and rapid fluctuations of the electronic spins on the ESR timescale (10^{-10} s) cause the narrowing of the ESR line (so-called exchange narrowing). The sharp decrease of the ESR line width is related to the slowing down of the rotational motion of the C_{60} units, as also observed by ^{13}C-NMR spectroscopy [19]. Therefore T_O is the orientational ordering temperature of the C_{60}^- ions. In this sense, TDAE-C_{60} is a highly interesting material as the orientational ordering of C_{60} units directly influences the magnetic interactions [20, 21] through the different relative orientations of the molecular orbitals and different overlap of the neighbouring electronic wave functions. This relationship between the orientational and spin degrees of freedom has been subsequently confirmed by the isolation and characterisation of two structural modifications of TDAE-C_{60} single crystals, designated as α- (ferromagnetic) and α'- (non-magnetic) phases [22, 23]. The two structural forms essentially differ only in the degree of the orientational order of the C_{60}^- ions [24]. Annealing of the α'-form at or above room temperature leads to a transformation to the ferromagnetic α-modification.

Orientational disorder effects, associated with the quasi-spherical nature of the fullerene molecules are abundant in C_{60}-based compounds. Orientational order in pristine C_{60} can be understood in terms of the combination of van der Waals and electrostatic interactions which result in the most electron-poor regions (the pentagonal faces) of the molecules facing the most electron-rich regions (the higher bond-order 6:6 bonds) of adjacent molecules [25].

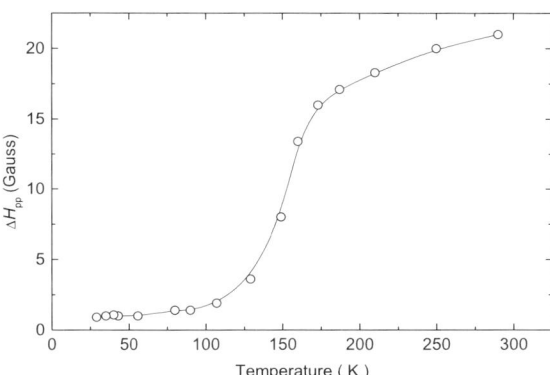

Fig. 9. Temperature dependence of the ESR (9.6 GHz) line width in TDAE-C_{60} single crystal. The solid line is a guide to the eye

However, disorder is still present, as a second "contact motif" is also present involving hexagons facing 6:6 bonds of neighbouring molecules [103]. Although the orientational potential in TDAE-C_{60} is much more complicated, we expect that similar structural motifs should be found and for optimal contact between adjacent C_{60}^- ions the ferromagnetic interactions are favoured. In fact in a recent study, Kambe et al. [26] studied in detail the effects of annealing on the magnetic and structural properties. It was found that well-annealed samples undergo a structural phase transition at around 180 K – most probably associated with the C_{60}^- orientational ordering previously detected by ^{13}C-NMR – and that in this case the saturated magnetisation increases to 0.9(1) μ_B/C_{60}^- ion. On the other hand, the as-grown crystals undergo no structural phase transition at least down to 30 K and do not order ferromagnetically. This underlines the importance of orientational order/disorder effects on the magnetic properties of TDAE-C_{60} compound. A theoretical model has been proposed in which the effective ferromagnetic exchange coupling strength, J_{eff}, shows pronounced variation with varying relative orientation of the C_{60}^- ions [21]. This model naturally explains the controversy raised by a number of contradictory reports on the magnetic properties.

Magnetic resonance experiments on magnetically ordered phases differ from ordinary ESR on paramagnetic samples in which only one electronic spin interacting with weak local fields is excited. The presence of large exchange fields in magnetic systems, typically orders of magnitude larger that the applied field, causes a coherent precession of all the electronic spins. The entire magnetisation of the sample is thus excited and the precession frequency depends on the magnitudes of the exchange, anisotropy, and demagnetisation fields. The relationship between the resonance frequency and the resonance field becomes strongly non-linear and depends on the nature of the magnetic ordering [27]. While in simple ferromagnets only one resonance mode is predicted theoretically, in antiferromagnets and weak ferromagnets two resonance modes should be found. For instance, for both uniaxial and weak ferromagnets, when the external magnetic field is perpendicular to the easy axis, a resonance mode with a dip in the resonance field-resonance frequency relation at a resonance field equal to the anisotropy field is found. However, in weak ferromagnets, an additional high-frequency antiferromagnetic type mode is also present.

On the other hand, the magnetic resonance in spin-glasses is very complicated and strongly dependent on thermal history [28]. For instance, due to hysteresis effects, the resonance line position under field-cooling conditions depends on whether the measured field is parallel or antiparallel to the cooling field. Superparamagnetic materials [29] behave in a similar fashion to normal paramagnets and the resonance position of the observed lines depends linearly on the resonance frequency. As a result, the (anti)ferromagnetic resonance technique provides an extremely sensitive tool for the determination of the ground state and the microscopic parameters in a variety of magnetic materials. The theoretical curves depicting the resonance frequency-resonance field relationship in ferromagnets, antiferromagnets and weak ferromagnets are summarised in Fig. 10.

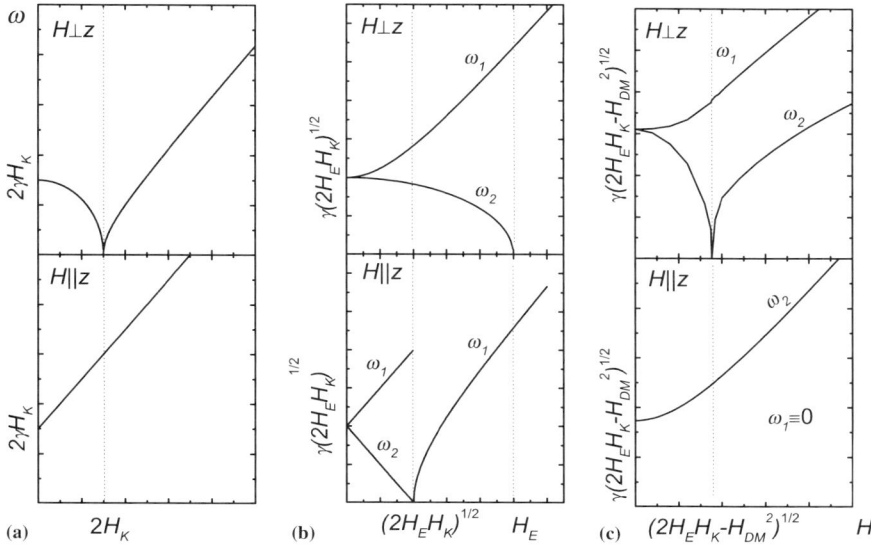

Fig. 10. Resonance frequency-resonance field relationship for (**a**) ferromagnets, (**b**) antiferromagnets, and (**c**) weak ferromagnets. Here H_E is the exchange field, H_K is the anisotropy field, and H_{DM} is the Dzyloshinskii-Moriya field

Clear evidence for the presence of a ferromagnetic resonance in TDAE-C_{60} was found in the radiofrequency region [30]. A single resonant line was observed in high-frequency experiments performed between 110 MHz and 220 GHz [31]. Below 110 MHz, a new line emerges at zero field which shifts to higher resonant fields with decreasing resonant frequency. The two resonant lines, high and low field, merge together below 50 MHz. The dependence of the resonant frequency on the resonant field is strongly non-linear and shows a characteristic dip at $H = 30$ G (Fig. 11), as expected for both uniaxial and weak ferromagnets (Fig. 10). Since no other resonant modes are observed at higher frequencies, the possibility of a weak ferromagnetic ground state in TDAE-C_{60} can be ruled out. The strongly non-linear behaviour (Fig. 11a) also eliminates the possibility of superparamagnetism and leads to the description of the magnetic ground state in TDAE-C_{60} single crystals as that of a normal Heisenberg ferromagnet. Agreement between theory and experiment becomes quantitative, if the demagnetising field effects are also taken into account. This leads to values of the anisotropy field, $H_K = 29$ G and the demagnetising field, $H_{dem} = -39$ G. The non-linear dependence of the resonance field on the resonance frequency disappears above the transition temperature in the paramagnetic phase (Fig. 11b).

In conclusion, we stress that TDAE-C_{60} is thus far the only authenticated ferromagnetic material based on C_{60}. Despite considerable efforts, doping with other organic donors invariably leads only to paramagnetic materials. It appears that the fortuitous combination of a number of factors, including

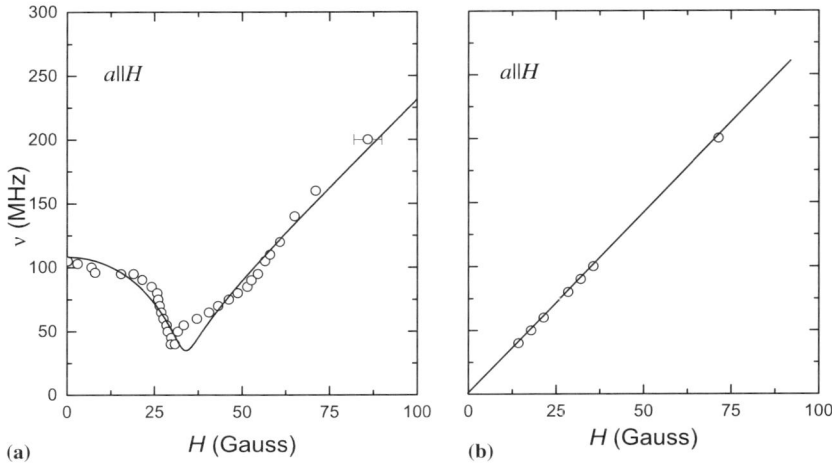

Fig. 11. Dependence of the resonance frequency on the resonance field ($a \| H$) in TDAE-C_{60} single crystal: (**a**) at $T = 5$ K ($<T_C$) and (**b**) at $T = 20$ K ($>T_C$)

orientational ordering of C_{60}^- ions, short C_{60}-C_{60} distances along the c-axis, and the role of TDAE$^+$ ions results in the formation of a ferromagnet with such an unusually high Curie temperature, unique among currently known non-polymeric, purely organic systems.

3
Antiferromagnetic Ordering in (NH$_3$)K$_3$C$_{60}$

Intercalation of solid C_{60} with alkali metals affords a series of salts with composition A_xC_{60} (A = alkali metal, $0 < x \leq 6$). Compounds with stoichiometry A_3C_{60} are invariably metallic, becoming superconducting with critical temperatures, T_C as high as 33 K at ambient pressure [32] with T_C increasing monotonically as the interfullerene separation, d increases [5, 33]. This can be rationalised in terms of increasing density-of-states at the Fermi level, $N(\varepsilon_F)$ with increasing d, resulting from the decrease in the overlap between the molecules that leads to band narrowing. As a consequence, in order to obtain high-T_C fullerides, large interfullerene spacings are needed. Ammoniation has proven an excellent method to achieve large expansions of the unit cells of fulleride salts, as neutral NH$_3$ molecules co-ordinate to the alkali ions, leading to large effective radii for the resulting $(NH_3)_x^{A+}$ species. Successful ammoniation of superconducting Na$_2$CsC$_{60}$ (T_C = 12 K) has afforded the cubic (NH$_3$)$_4$Na$_2$CsC$_{60}$ salt with T_C of 29.6 K [34]. On the other hand, reaction of K$_3$C$_{60}$ (T_C = 19 K) with NH$_3$ led to the isolation of the orthorhombic (NH$_3$)K$_3$C$_{60}$ phase which is non-superconducting [35]. However, application of a pressure >1 GPa leads to recovery of superconductivity in (NH$_3$)K$_3$C$_{60}$ with T_C = 28 K [36].

Structural characterisation [35, 37] of $(NH_3)K_3C_{60}$ at room temperature revealed an anisotropic expansion of the fulleride array leading to a face-centred orthorhombic cell with lattice constants $a = 14.917(10)$ Å, $b = 14.971(10)$ Å, $c = 13.692(4)$ Å (space group $Fmmm$). The observed increase in the interball separation and the reduction in crystal symmetry should have important consequences for electron localisation and the loss of superconductivity. Below 150 K, a phase transition occurs to another orthorhombic structure (space group $Fddd$) in which the unit cell dimensions double along all three crystallographic directions [37]. The origin of the observed superstructure was originally identified with the antiferroelectric ordering of the K^+-NH_3 pairs, which reside in the octahedral sites [37]. Its signature was also evident in neutron inelastic scattering measurements in which a librational mode at ∼6.5 meV, implying freezing of the ammonia rotation, appears below 150 K [38]. However, recent neutron diffraction experiments have revealed additional subtleties in the adopted structure arising from orientational ordering of the C_{60} units, which adopt two different orientations related by a 90° rotation about the c orthorhombic axis and order along the a axis (antiferrorotative ordering, Fig. 12) [39]. Structural work at elevated pressures also showed that above ∼1 GPa the system is isostructural with the ambient-pressure low-temperature phase [39]. With increasing pressure, $(NH_3)K_3C_{60}$ becomes increasingly more anisotropic with an increased size difference between the lattice constants a and b.

The position of the ^{13}C-NMR line (195 ± 2 ppm) in $(NH_3)K_3C_{60}$ [40, 41] falls within the range of other C_{60}^{n-} ions and is only slightly larger than that of

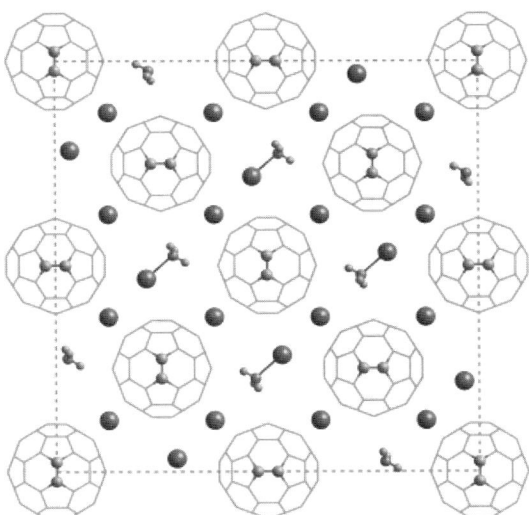

Fig. 12. Low-temperature orthorhombic crystal structure of $(NH_3)K_3C_{60}$ (space group $Fddd$), projected on the [110] plane. The observed superstructure arises from the antiferrorotative order of the C_{60} units and the antiferroelectric order of the K^+-NH_3 pairs

K_3C_{60} (187 ppm). The narrow ^{13}C-NMR line at room temperature implies quasi-isotropic fast reorientation on the NMR timescale of the C_{60}^{3-} ions. The ^{13}C-NMR spectra broaden below 150 K due to the freezing out of the molecular motions. C_{60}^{3-} molecular dynamics has been studied in detail by the low-energy inelastic neutron scattering measurements as a function of the scattering vector Q and in the temperature range between 310 K and 30 K [42]. The C_{60}^{3-} ions were found to execute small-amplitude librations about their equilibrium positions, giving rise to well-defined peaks near 3.1 meV at low temperature (Fig. 13). The librational energy is smaller than in the parent K_3C_{60} salt (4.04 meV at 12 K), and the peaks are much broader, reflecting a weaker and more anisotropic orientational potential than that encountered in the parent K_3C_{60} [42].

The electronic and magnetic properties of $(NH_3)K_3C_{60}$ were measured with the NMR, ESR, and μSR techniques. Measurements of ^{13}C spin-lattice relaxation rate, $1/T_1$, showed that $(NH_3)K_3C_{60}$ is metallic above 40 K and the Korringa law is followed between 40 and 100 K. However below 40 K,

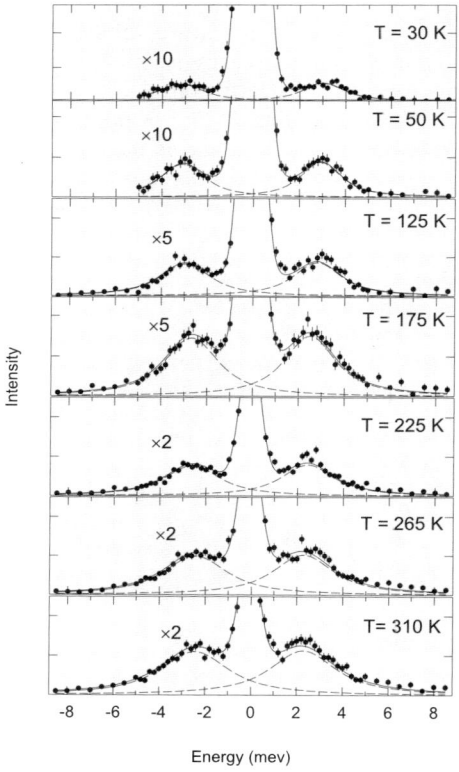

Fig. 13. Representative NIS spectra of $(ND_3)K_3C_{60}$ in the temperature range 30–310 K at constant $Q = 5.5$ Å^{-1}

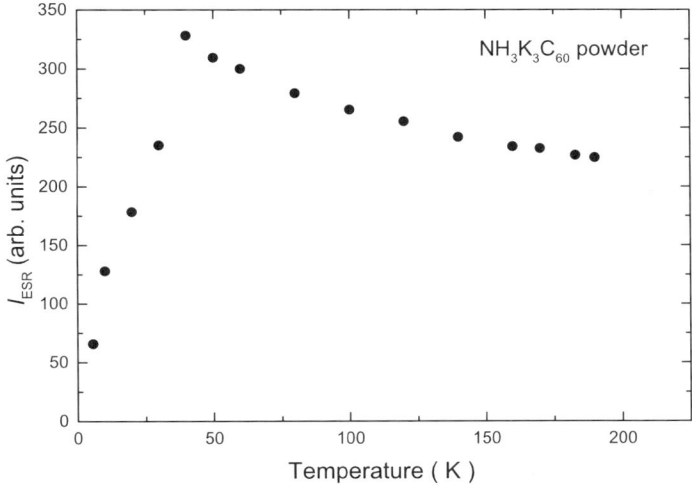

Fig. 14. Temperature dependence of the ESR intensity in $NH_3K_3C_{60}$

$1/T_1$ suddenly drops corroborating the occurrence of a phase transition [40, 41].

More information about the low temperature transition at 40 K can be deduced from the ESR results which showed that $(NH_3)K_3C_{60}$ is a narrow band metal [40, 41, 43, 44]. The room temperature spin susceptibility, χ_S determined by double integration of the integrated derivative ESR spectra, is 5.0×10^{-4} emu/mol. The value of χ_S is characteristic for the metallic state of fulleride salts and is comparable with that measured in K_3C_{60} [45]. On cooling, the ESR spin susceptibility first slightly increases down to 40 K (Fig. 14). The increase in χ_S is rather exceptional for the A_3C_{60} systems, where the susceptibility normally decreases due to the lattice contraction and thus the reduction of the density of states at the Fermi level [46]. The variation of χ_S with temperature is difficult to explain simply in terms of metallic or localised spin behaviour. The increase of the susceptibility is too weak to originate from Curie–Weiss dependence [47]. However, the same type of behaviour was found in polymeric RbC_{60} and has been ascribed to the strong antiferromagnetic correlations in a narrow band metal system. Below 40 K, the ESR susceptibility suddenly drops by more than an order of magnitude. Such a drop in χ_S could be a result of antiferromagnetic ordering, transition to a spin-density wave (SDW) or even to non-magnetic charge-density wave (CDW). It is of interest to note that the ESR line width, which should increase in the vicinity of the transition due to the increased spectral density of the local field fluctuations, does not show any anomaly. If the ground state is characterised by long range ordering of electronic spins, antiferromagnetic or SDW type, then antiferromagnetic resonance should be detected in high-frequency ESR experiments. Magnetic resonance experiments at frequencies between 9 and 225 GHz [44, 47] indeed identified an additional resonance line, which has been attributed to the antiferromagnetic resonance. Below 40 K, the resonance line broadens and shifts as the magnetic order develops. For the antiferromagnetic resonance in powdered samples, the line width, ΔH_{pp} should decrease with increasing resonance field, H_0 ($H_0 \gg H_{SF}$, where H_{SF} is the spin-flop field) according to the relationship: $\Delta H_{pp} \propto H_{SF}^2/H_0$. However, while ΔH_{pp} between 9 GHz ($H_0 = 0.3$ T) and 75 GHz ($H_0 = 2.7$ T) narrows, for frequencies between 75 GHz and 225 GHz, the line width depends only weakly on external field and does not follow the $1/H_0$ field dependence. One of the reasons could be that the line width narrowing is compensated by the g-factor anisotropy broadening effects [47]. Experiments on single crystals of $NH_3K_3C_{60}$ should certainly help to clarify the nature of the observed line.

Even though the early ESR and NMR measurements of the electronic and magnetic properties have shown that $(NH_3)K_3C_{60}$ is a narrow band metal which exhibits a transition to an insulating ground state at about 40 K, the nature of the low temperature state remained controversial as the experimental results were unable to distinguish between a magnetic or a non-magnetic ground state. The problem of the existence of non-zero internal local magnetic fields at low temperature in $(NH_3)K_3C_{60}$ and its deuterated analogue, $(ND_3)K_3C_{60}$ was addressed by using 100% spin polarised positive muons (μ^+) in the absence of external fields [48, 49]. At temperatures higher than 40 K, only a non-oscillating signal is evident. In this temperature range, the μ^+ spin relaxation is determined by the fluctuations of the nuclear moments, which appear frozen into a disordered spin configuration. In addition, the μ^+ spin is relaxed by the rapidly fluctuating electron spins, which give rise to the Lorentzian relaxation component with $\lambda = 0.030(2)$ μs^{-1} at 100 K. Cooling-

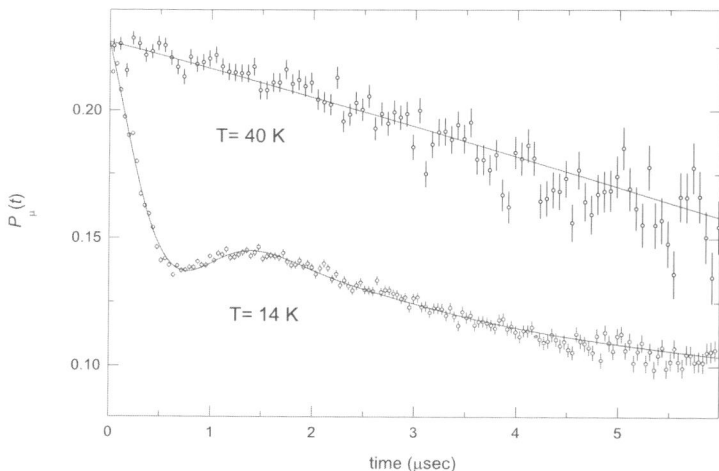

Fig. 15. Evolution of the zero-field (ZF) μ^+ spin polarisation, $P_\mu(t)$ at 14 and 40 K for $(ND_3)K_3C_{60}$. The solid lines through the data are fits to the functions of Eqs. 1 and 2, respectively

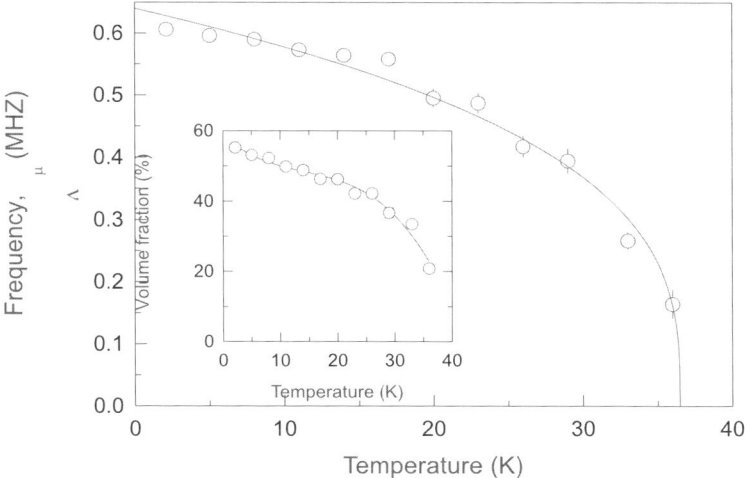

Fig. 16. Temperature dependence of the ZF muon precession frequency, ν_μ (open circles) in $(ND_3)K_3C_{60}$; the solid line is a fit of the data to the critical law, Eq. 2. The inset shows the temperature dependence of the volume fraction (open circles) of the magnetically ordered component in $(ND_3)K_3C_{60}$; the broken line represents a guide to the eye

down leads to a slowing-down of the electron spin dynamics within the paramagnetic domains with λ reaching a value of 0.046(1) μs^{-1} at 40 K [49].

Below 40 K (Fig. 15), the shape of the time-dependent μ^+SR spectra changes, as a short-lived oscillating component, whose depolarisation gradually increases with decreasing temperature, appears. The observation of a precession signal in zero external field is clear evidence of the existence of coherent ordering of the electronic spins and indicates unambiguously the onset of long-range antiferromagnetic order. The precession frequency, ν_μ is 0.564(4) MHz at 14 K, corresponding to a static local field at the muon site, $\langle B_\mu \rangle = 41.6(3)$ G. In addition, the depolarisation rate, λ has a value of 2.14(4) μs^{-1} at 14 K, implying a distribution of local fields with a width $\langle \Delta B^2 \rangle^{1/2} = 25.1(4)$ G, only smaller than $\langle B_\mu \rangle$ by a factor of 1.7. The muons thus experience a local field with large spatial inhomogeneities, which may be due to a number of physical reasons, including orientational disorder effects of the fullerene molecules. The temperature evolution of the muon frequency, ν_μ, is described by the equation: $\nu_\mu = \nu_0[1 - (T/T_N)]^\beta$, with $\nu_0 = 0.64(1)$ MHz, $T_N = 36.5(3)$ K, and $\beta = 0.32(3)$ (Fig. 16). The critical exponent, β, is close to the value expected for a conventional 3D Heisenberg antiferromagnet ($\beta = 0.36$). As the μ^+ site is not known for a powder sample, a value of the magnetic moment cannot be extracted directly. However, we note that ν_0 is 1.4 times smaller than the frequency observed in $(TDAE)C_{60}$, in which the magnetic moment is ~ 1 μ_B/molecule [26]. Assuming similar stopping sites for the two samples, we find $\mu(T = 0 \text{ K})$ is in the order of 0.7 μ_B/molecule.

Data collected at 14 K in a 200 G longitudinal field reveal a complete recovery of the asymmetry. As the effect of applied longitudinal fields is to allow the depolarisation due to dynamic or fluctuating moments to be decoupled from that due to static components, we conclude that the origin of the observed relaxation in zero field is of quasi-static nature.

Recent ^{13}C-, ^{1}H-, and ^{39}K-NMR measurements have also confirmed the antiferromagnetic nature of the low-temperature ground state in $(NH_3)K_3C_{60}$ [50]. However, in disagreement with earlier ESR and NMR results, the authors conclude that that even above 40 K, the system can be described as an $S = 1/2$ localised spin system, and not as a highly-correlated metal.

In conclusion, $(NH_3)K_3C_{60}$ and its deuterated analogue show a transition below $\sim$37 K to a long-range ordered antiferromagnetic state, characterised by considerable spatially inhomogeneous effects. The suppression of superconductivity is thus associated with effects of magnetic origin, providing an important analogy with the well-established phenomenology in high-T_C and organic superconductors. Thus, the increased interfullerene separation together with the symmetry reduction in $(NH_3)K_3C_{60}$ have important consequences for the electronic properties of the material. The overlap between the molecules decreases substantially, leading to a reduced bandwidth, W, and an increased (U/W) ratio, which for fixed band filling drive the system to an antiferromagnetic Mott insulating state [51]. Then the recovery of superconductivity at pressures higher than 1 GPa ($T_C = 28$ K) [36, 39] should presumably be associated with the suppression of the magnetic transition. Fig. 17 displays a schematic electronic phase diagram for C_{60}^{3-} compounds as a function of the volume per C_{60}^{3-} anion [or equivalently increasing (U/W) ratio].

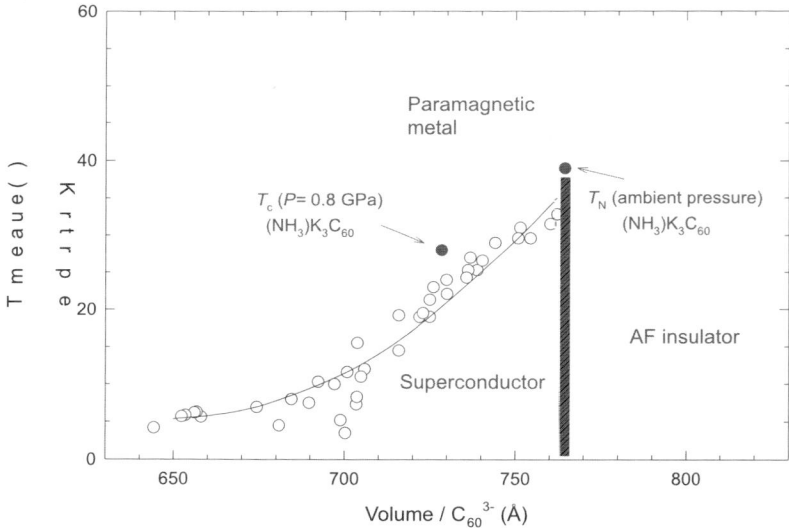

Fig. 17. Schematic electronic phase diagram of C_{60}^{3-} compounds, showing the approximate location of the metal (superconductor)-insulator phase boundary. The open symbols are literature values of T_C for a variety of superconducting fullerides, while the solid symbols mark T_N (ambient pressure) and T_C (>1 GPa) of $(NH_3)K_3C_{60}$

Future studies of the electronic properties of systems lying on either side of the metal (superconductor)-insulator phase boundary promise to lead to intriguing results.

4
Magnetism in Polymeric Fullerides

4.1
Magnetic Properties of the AC_{60} Family

Stable phases at the $x = 1$ point in the phase diagram of A_xC_{60} (A = K, Rb, Cs) were not identified until some time after the superconducting $x = 3$ and insulating $x = 4$ and 6 phases were discovered and characterised [52]. The first indications of stable $x = 1$ phases came from Raman [53], photoemission [54], X-ray diffraction [55], and ESR experiments [56]. Soon afterwards it was observed that RbC_{60} can be either conducting or insulating below room temperature, depending on the thermal history. Depending on the cooling protocols used, a number of distinct phases has now been stabilised (Fig. 18).

At high temperature (above 400 K), AC_{60} is fcc with alkali ions residing in the octahedral sites [55]. Below 400 K, the most stable form was originally shown to adopt orthorhombic crystal symmetry [57, 58] with pronounced quasi-one-dimensional character and unusually close contacts between the fulleride ions; partial covalent bonding between fullerenes along the chains and their consequent molecular deformation from icosahedral symmetry have been postulated to account for these observations. Structural characterisation was originally achieved by synchrotron X-ray powder diffraction [58] studies, which established a cell of dimensions, $a = 9.11$ Å, $b = 10.10$ Å, and $c = 14.21$ Å (space group $Pmnn$) for RbC_{60} at room temperature with fulleride linkages along the polymer chains achieved by a [2 + 2]cycloaddition mechanism, in analogy with the photopolymerisation reaction of pristine C_{60}. The contraction along the polymerisation a axis is $\sim$0.9 Å, compared to the centre-to-centre interfullerene distances of about 10 Å, encountered in monomeric fullerenes and their derivatives. However, subsequent structural work on KC_{60} and AC_{60} (A = Rb, Cs) [59, 60] revealed that they possess different chain orientations about their axes, which are described by distinct

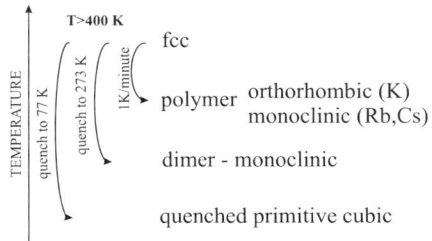

Fig. 18. Sequences of phases observed in AC_{60} fullerides as a function of temperature and cooling rates

space groups $Pmnn$ (orthorhombic) and $I2/m$ (monoclinic), respectively (Fig. 19).

A medium quenching rate to just below room temperature leads to the stabilisation of a low-symmetry insulating structure which has been characterised by synchrotron X-ray diffraction studies to comprise $(C_{60}^-)_2$ dimers, bridged by single C—C bonds [61]. The RbC_{60} dimer phase crystallises in the monoclinic space group $P2_1/a$ with lattice parameters $a = 17.141(5)$ Å, $b = 9.929(5)$ Å, $c = 19.277(5)$ Å, $\beta = 124.4°$ at 220 K (Fig. 20) and transforms back to the polymer phase on heating. The chain axis is now along b and is somewhat longer than that found in the RbC_{60} polymer, implying a smaller contraction of ~ 0.7 Å from the interfullerene separations found in monomeric fullerenes. This is consistent with the presence of a single bridging C—C bond, instead of the four-membered rings of the polymers. Another metastable conducting phase of RbC_{60} and CsC_{60} was reported with very rapid quenching by immersing the high-temperature sample in liquid nitrogen [62]. In this case, the 3D isotropic character of the structure is retained while orientational ordering of the C_{60}^- ions leads to a simple cubic structure (space group $Pa\bar{3}$) with $a = 13.9671(3)$ Å at 4.5 K [63].

The AC_{60} (A = K, Rb, Cs) fullerides have attracted particular interest not only because of their rich structural phase diagram but also for their electronic, conducting, and magnetic properties which have remained the subject of controversy for a while [64]. A metal-insulator transition below 50 K is accompanied by the stabilisation of a magnetic state in the monoclinic phase of RbC_{60} and CsC_{60}, whereas orthorhombic KC_{60} remains metallic to low

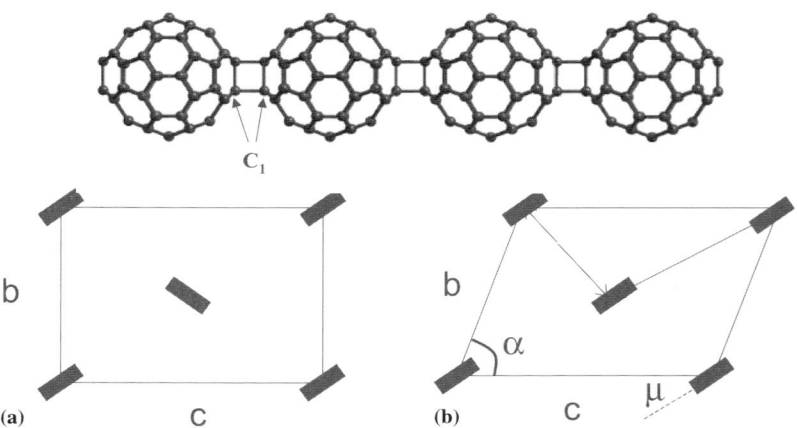

Fig. 19. Linear polymer chains formed by [2 + 2]cycloaddition in AC_{60} salts. Schematic drawing of chain orientations for (a) $Pmnn$ space group (KC_{60}) and (b) $I2/m$ space group (RbC_{60}, CsC_{60}). The shaded bars indicate the orientation of covalently bonded polymer chains along the a axis by projection on the crystallographic bc plane. C_1 labels the bridging carbon atom, α is the monoclinic angle, and μ is the angle between the cycloaddition planes and the c axis (μ = 510, 470, and 460 for KC_{60}, RbC_{60}, and CsC_{60}, respectively)

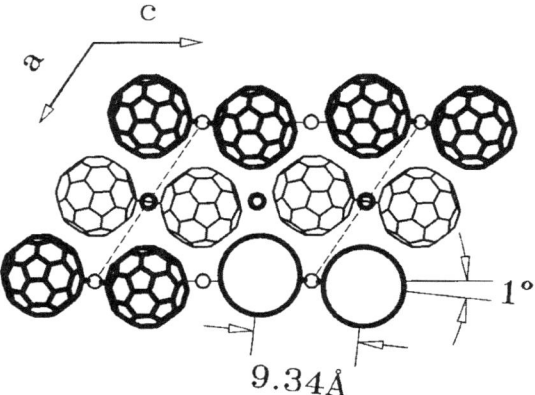

Fig. 20. Structure of dimerised phase of RbC_{60} stabilised by medium quenching from high temperature to below room temperature

temperatures [65]. The nature of the magnetic transition has been controversial, as both quasi-one-dimensional electronic instabilities [57, 64–68] and three-dimensional magnetic ordering [69, 70] have been proposed. According to density-functional calculations [69], the rehybridisation of carbon orbitals at the C_{60}—C_{60} bonds impedes conduction along the chains and leads to strongly three-dimensional electronic properties with a 3D antiferromagnetically ordered ground state. On the other hand, in the original report on the magnetic properties of polymeric RbC_{60} [57], it was proposed that it behaves as a quasi-one dimensional metal. Measurements of the ESR susceptibility find a large and approximately Pauli-like spin susceptibility ($\chi_S \approx 6 \times 10^{-4}$ emu/mol, Fig. 21a). The temperature evolution of the static susceptibility, χ_S, as measured by EPR, revealed a sharp decrease to zero in the vicinity of 50 K, attributed to a spin density wave (SDW) instability of the conducting C_{60}^- linear chains [57]. Antiferromagnetic correlations are also supported by the temperature dependence of the ESR line width (Fig. 21b), the g-factor of the ESR line (Fig. 21c) [57] as well as the enhanced ^{13}C-NMR spin-lattice relaxation [71]. Zero-field muon spin relaxation (ZF-μSR) studies of RbC_{60} and CsC_{60} samples [72–74] revealed frozen electronic moments in the low-temperature phase, but spontaneous Larmor frequencies which would have unambiguously implied long-range magnetic order (LRO), associated with antiferromagnetism (AF) or spin density wave (SDW) formation, were absent. The muons experienced a local magnetic field that peaked close to zero and with large spatial inhomogeneities which may be due to a number of physical factors, including a wide distribution of chain $(C_{60}^-)_n$ lengths and orientational disorder effects.

In order to clarify the true nature of the long-range magnetic order in the AC_{60} polymers, antiferromagnetic resonance should be observed. Unfortunately the lack of single crystalline samples makes it necessary to rely on powder data. The first report of the observation of antiferromagnetic

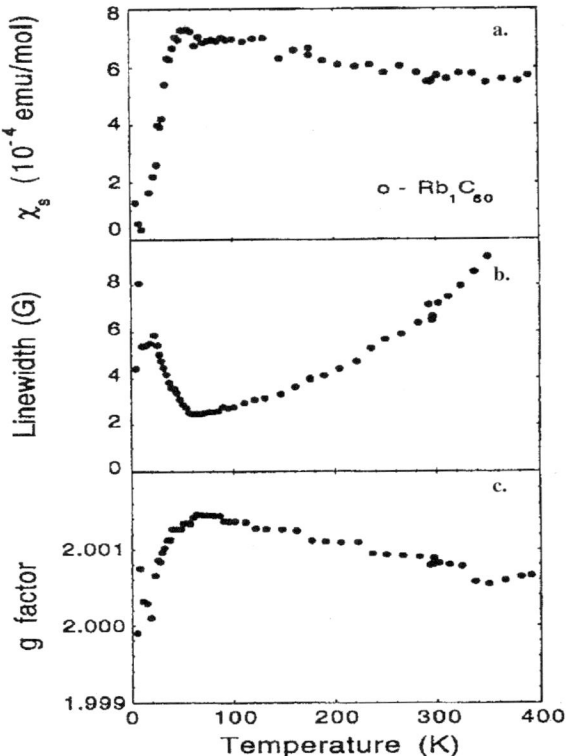

Fig. 21. Temperature dependence of the ESR (**a**) susceptibility, (**b**) line width, and (**c**) g-factor measured in polymeric RbC_{60}

resonance came from Jannosy et al. [68] who measured the ESR spectra of RbC_{60} at high frequencies (75 GHz, 150 GHz, and 225 GHz). At these frequencies, a broad line whose frequency is inversely proportional to the resonance field has been observed [64, 68]. The sublattice magnetisation was found to be independent of applied magnetic field up to at least 8 T, while magnetic fluctuations were present between 35 and 50 K. Comparison with the SDW system, $(TMTSeF)_2PF_6$, implied that the fulleride polymers are also quasi-1D SDW systems with 3D ordering at low temperatures. However, as discussed by Bennati et al. [64], unequivocal proof for an antiferromagnetic ground state cannot be established, as the data could be also interpreted in terms of a spin-glass model and final confirmation should await single-crystal AFMR studies. If the antiferromagnetic ground state scenario is invoked, a Néel temperature, $T_N = 25$ K, and a spin flop field of about 0.34 T can be estimated.

The low temperature state of the CsC_{60} polymer seems to be even more complicated. Recent ^{13}C- and ^{133}Cs-NMR measurements [75] have detected the appearance of a spin-singlet (non-magnetic) ground state below $T_S = 13.8$ K which coexists with the magnetic order that develops at T_N. The application of pressure first suppresses the magnetic order and a homogeneous non-magnetic ground state is stabilised at 5 kbar. It was proposed that the development of the non-magnetic phase might be correlated either with a structural change, tentatively ascribed to the occurrence of a spin-Peierls transition, or with an electronic instability. Synchrotron X-ray diffraction measurements have identified the appearance of a spontaneous thermal contraction below 14 K which provides the signature of magnetoelastic coupling and could trigger the occurrence of the spin-singlet ground state [60]. We note here that a similar situation is also encountered in the prototypical quasi-1D systems $(TMTTF)_2PF_6$ [76], $(BCPTTF)_2AsF_6$ [77], and $(TMTTF)_2Br$ [78] where the presence of spin-Peierls fluctuations has been recently identified. Spin-Peierls fluctuations were detected by X-ray diffraction and weak diffusive lines, located midway between the main Bragg reflections, were observed [79]. Similar X-ray diffraction experiments on the CsC_{60} polymer should be valuable in order to clarify further the nature of its low-temperature spin-singlet state.

A final comment concerns the spin susceptibilities measured by EPR for the 1D polymer and the quenched 3D CsC_{60} phases. The observed Pauli-like susceptibilities imply that both phases are conducting [62]. However, while the quasi-1D polymer phase shows a metal-insulator instability, the primitive cubic phase reveals contrasting electronic behaviour on the local scale, as revealed by ^{13}C- and ^{133}Cs-NMR measurements [80]. A partial spin gap opens below $T < 50$ K, ascribed to the occurrence of localised spin singlets on a small fraction of the C_{60}^- ions. ^{133}Cs-NMR shows that there are two inequivalent Cs sites (Fig. 22), despite the lack of any observed distortions from cubic symmetry in structural studies [63]. For one of these sites, the NMR shift and $(TT_1)^{-1}$ follow an activated law, confirming the existence of the spin gap.

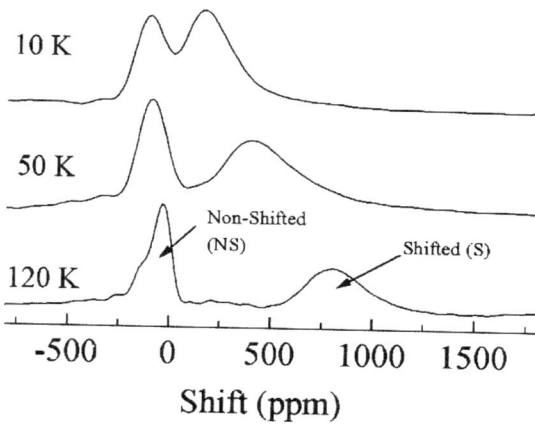

Fig. 22. ^{133}Cs-NMR spectrum of the quenched primitive cubic phase of CsC_{60}, showing the presence of two inequivalent Cs lines

4.2
Magnetic Properties of Na$_2$RbC$_{60}$ and Related Salts

Most prominent among alkali metal fullerides are the superconducting salts with composition A$_3$C$_{60}$. Geometrical considerations dictate that for large alkali ions (K$^+$, Rb$^+$, and Cs$^+$) occupying the small tetrahedral interstices (two per C$_{60}$ unit with a radius of 1.12 Å), the crystal structure of the A$_3$C$_{60}$ salts is face-centred cubic (*fcc*, space group $Fm\bar{3}m$) [81]. Rotational motion of the C$_{60}^{3-}$ ions is restricted and the adopted structure is understood in terms of the strong repulsive interactions between the alkali ions and the fulleride units. The latter are forced to expose towards the tetrahedral alkali ions the part of their quasi-spherical shape with the largest surface area, namely the hexagonal faces, thus leading to maximisation of the A$^+$-C(C$_{60}$) distances. The control of the structural properties of A$_3$C$_{60}$ by the tetrahedral ions is manifested in the expanded dimensions of the *fcc* unit cell ($a = 14.240$ Å for K$_3$C$_{60}$, $a = 14.157$ Å for C$_{60}$ at room temperature) and the extreme robustness of the structure. The superconducting transition temperature is affected in essentially an identical way by both physical and chemical pressure in these systems. The $N(E_F)$ is modulated through the changes of the electronic overlap between the C$_{60}^{3-}$ ions leading to the changes in the bandwidth.

The sensitive control of the properties of intercalated fullerides by the ion occupying the tetrahedral interstices is immediately manifested when Na$^+$ are introduced in the tetrahedral holes of a *ccp* array of C$_{60}^{3-}$ ions. As the Na$^+$ ionic radius (0.95 Å) is smaller than the size of the hole, there is evidently enough space for the C$_{60}^{3-}$ ions to rotate in such a way as to optimise both the attractive Na$^+$—C$_{60}^{3-}$ interactions and the C$_{60}^{3-}$—C$_{60}^{3-}$ contacts. The resulting structure for Na$_2$AC$_{60}$ (A = K, Rb, Cs) is found to be *fcc*, comprising quasi-spherically disordered C$_{60}^{3-}$ ions. On cooling, a phase transition occurs in the vicinity of room temperature to a primitive cubic structure (*pc*, space group $Pa\bar{3}$) [82, 83]. The primitive cubic family of Na$_2$(A,A')C$_{60}$ salts displays a much steeper rate of change of T_C with interfullerene spacing than that exhibited by *fcc* fullerides, while at the same time, the effects of physical [84] and chemical [85] pressure on the superconducting properties are not identical, and chemical pressure suppresses T_C much faster than physical pressure does. These peculiarities in the electronic properties of Na$_2$AC$_{60}$ were originally ascribed to the dependence of $N(E_F)$ on the orientational order of the C$_{60}$ molecules in the primitive cubic (*pc*) structure. However, this conjecture has proven too simplistic, as subsequent work [5, 86] revealed a much more complicated structural and electronic phase diagram for Na$_2$AC$_{60}$ (Fig. 23). Thus, it was found that, if special care was taken to cool the sample very slowly down to 180–200 K (with cooling rates as slow as 1 K/h), the ground state of Na$_2$RbC$_{60}$ below about 250 K was that of a quasi-one-dimensional polymeric phase with monoclinic crystal symmetry and with a short interball centre-to-centre distance of ~9.38 Å, which is essentially insensitive to cooling [86, 87].

The polymerisation of C$_{60}^{3-}$ ions in Na$_2$RbC$_{60}$ is characterised by a different structural motif than that encountered in the extensively studied RbC$_{60}$ polymer phase, involving the formation of a single C—C bridging bond

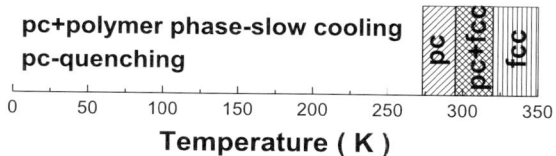

Fig. 23. Phase diagram of Na_2RbC_{60} at ambient pressure

(Fig. 24) [87, 88]. Partial replacement of Rb by Cs to give $Na_2Rb_{1-x}Cs_xC_{60}$ ($0 \leq x \leq 1$) allows the formation of a family of isostructural polymer phases in which the amount of polymer at low temperatures strongly depends on the Cs content, x, as well as the cooling rate [39, 89–91]. Structural studies have confirmed the occurrence of polymerisation for the entire $Na_2Rb_{1-x}Cs_xC_{60}$ family at both ambient and elevated pressure. Table 1 summarises the dependence of the interfullerene separation, d, in the primitive cubic precursor phase just below the onset of polymerisation on the Cs content, x [39]. The C_{60}–C_{60} interfullerene distance is sensitively controlled by x; for instance, it increases by $\sim$0.24% on going from Na_2RbC_{60} to $Na_2Rb_{0.3}Cs_{0.7}C_{60}$. This expansion is also reflected in the interchain separation of the corresponding polymer phases and is found to have dramatic consequences on the electronic and magnetic properties of the $Na_2Rb_{1-x}Cs_xC_{60}$ polymer phases. The interchain electron hopping is sensitively affected and increased quasi-one dimensional behaviour is encountered across the series with increasing x.

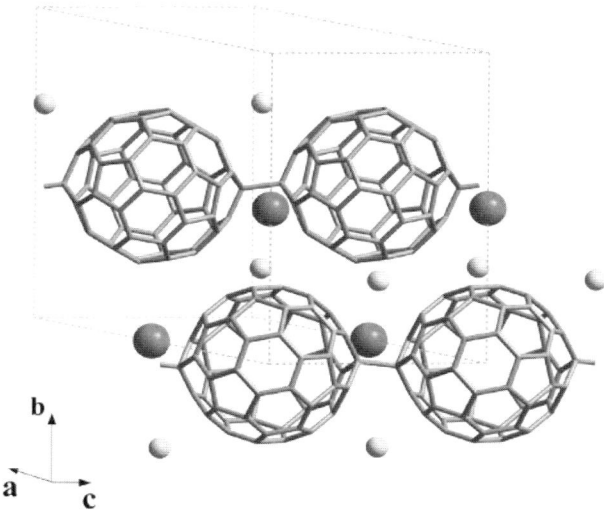

Fig. 24. Crystal structure of the Na_2RbC_{60} polymer. The polymer chains are aligned along the c axis of the cell with the fullerenes connected by single C—C bonds. Na^+ and Rb^+ ions are shown as small and large spheres, respectively

Table 1. Dependence of the interfullerene separation on Cs content for the primitive cubic series $Na_2Rb_{1-x}Cs_xC_{60}$ ($0 \leq x \leq 1$) at $\sim$250 K

Compound	d (Å)
Na_2CsC_{60}	9.9754(4)
$Na_2Rb_{0.3}Cs_{0.7}C_{60}$	9.9669(2)
$Na_2Rb_{0.5}Cs_{0.5}C_{60}$	9.9601(2)
$Na_2Rb_{0.8}Cs_{0.2}C_{60}$	9.9519(3)
Na_2RbC_{60}	9.9431(4)

The electronic properties of $Na_2Rb_{1-x}Cs_xC_{60}$ samples were investigated using the X-band (9.6 GHz) and high-field (110 GHz) ESR techniques [92–94]. The ESR spectra at low temperatures when both the cubic monomer and the polymer phases co-exist are composed of two lines: a broad line, characteristic for the cubic superconducting phase and a narrow component, originating from the polymeric phase (Fig. 25). Deconvolution of the X-band ESR spectra thus enables one to determine the temperature dependence of the spin susceptibility of the polymeric phase.

The temperature dependence of the intensities of the two components in Na_2RbC_{60} (Fig. 26) clearly reflects the changes in the spectra: the intensity of the narrow component gradually increases below $\sim$265 K at the expense of the high-temperature broad cubic component. However, the difference in the line widths of the two components is not as large as one would expect considering the one-dimensional versus three-dimensional structure of the polymeric and cubic phases, respectively. This implies that polymerisation in Na_2RbC_{60} results only in a partially reduced dimensionality of the electronic structure. It also explains the temperature dependence of the susceptibility of the polymeric phase, which decreases with decreasing temperature. The Na_2RbC_{60} polymer remains metallic down to 4 K without any typical instability, characteristic of low-dimensional metallic systems. In fact, the decrease in the ESR susceptibility with decreasing temperature could well result from lattice contraction, suggesting a 3D-like electronic structure [92]. It also suggests the absence of any magnetic correlations in this system and explains the fact that no low-temperature instability has been found in this compound.

An investigation of the electronic properties of the series of $Na_2Rb_{1-x}Cs_xC_{60}$ polymers with the X-band and high-field ESR (HF ESR) techniques allowed a systematic determination of the low-temperature magnetic properties. A very sensitive way of looking at the electronic dimensionality of these systems is the electron spin-lattice relaxation time, T_1 [95]. In the 3D metallic regime, the electron spin-spin relaxation time, T_2 is expected to become equal to T_1 and the measured homogeneous line width of the ESR spectrum is given by: $T_1 \approx T_2 = (2\hbar)/(g\mu_B \Delta H_{1/2})$. It has been shown before for A_3C_{60} cubic systems that T_1 is principally determined by the spin-orbit coupling of the conduction electrons to the lattice [96]. In this case, the spin-lattice relaxation time can be expressed by the Elliot expression [97]:

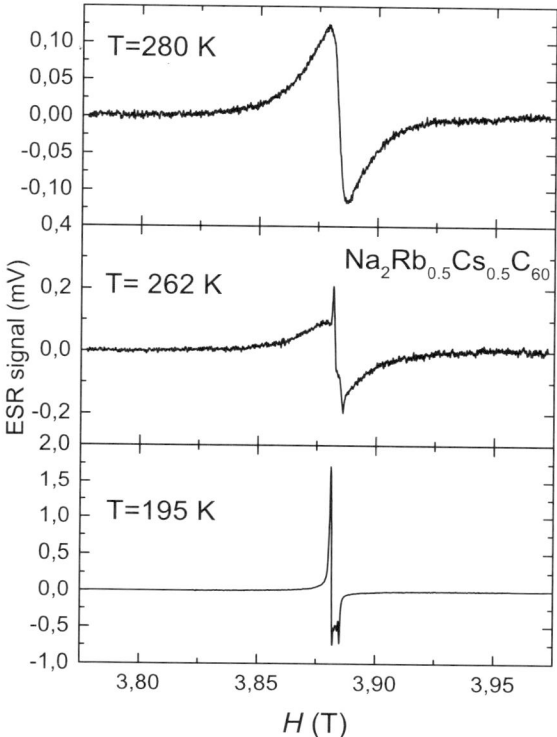

Fig. 25. Representative temperature evolution of the high-field (110 GHz) ESR spectra in $Na_2Rb_{1-x}Cs_xC_{60}$, showing the appearance on cooling of the narrow component, attributed to the polymeric phase, superimposed on the broad cubic component. The data shown here are for the sample with $x = 0.5$

$$T_1 = \frac{\alpha}{(\Delta g)^2}\tau_\| \quad T \gg \theta_D \tag{1a}$$

and

$$T_1 \approx \frac{\tau_\|}{(\Delta g)^2}\left(\frac{T}{\theta_D}\right)^2 \quad T \ll \theta_D \tag{1b}$$

where Δg is the shift of the g-factor from the free electron value, $g_e = 2.0023$, $\tau_\|$ is the relaxation time of the electric conductivity, θ_D is the Debye temperature, and α is constant, which for isotropic metals is on the order of 1 at high temperatures. It should be noted that even Elliot in his original contribution warned about the crudity of the calculations and that numerical factors involved in the equations above should be treated with caution. Tanigaki et al. [97] connected T_1 with the spin-orbit coupling constant, λ, of the intercalated alkali ions, introducing the expression, $\Delta g \approx \lambda/E$, where E is the energy separation of the considered valence states. This leads to

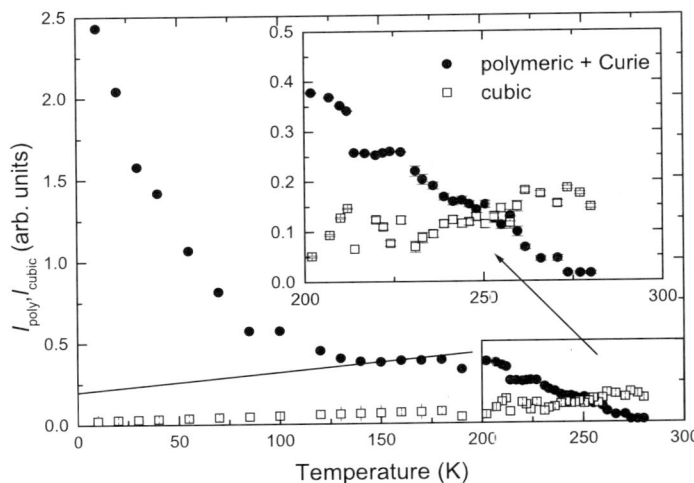

Fig. 26. Temperature dependence of the high-field ESR (110 GHz) intensity of the cubic (open squares) and polymeric (closed circles) components in Na_2RbC_{60}. The inset shows an expanded version of the diagram in the temperature range 200–300 K

$$\Delta H_{1/2}^{LS} = \frac{2\hbar}{g\mu_B} \frac{(\Delta g)^2}{\alpha \tau_\parallel} \propto \left(\frac{\lambda}{E}\right)^2 \tag{2}$$

and explains why the ESR line width of the cubic line increases on moving from K to Rb to Cs. The reported λ values of the alkali metals are: 0.3 cm^{-1} for Li, 17 cm^{-1} for Na, 58 cm^{-1} for K, 6820 cm^{-1} for Rb, and 9810 cm^{-1} for Cs. However, in anisotropic systems, α deviates from unity and should be estimated theoretically or experimentally. In a quasi-one-dimensional metallic system, one should also consider a cut-off of the 1D correlation of electronic motion due to the probability of escape from the 1D chain. The relaxation time, $\tau_\perp$, thus measures the mean lifetime of the 1D correlation. Experimental work on the prototypical quasi-one-dimensional system, TTF-TCNQ, has shown [98] that α could be expressed as: $\alpha \approx \tau_\perp/\tau_\parallel$. Increasing the degree of quasi-one-dimensionality by expanding the chain-chain separation in a conducting polymeric chain results in a reduced transverse transfer integral. This in turn will effectively increase α and result in line narrowing.

In Fig. 27 we show the dependence of the high-field ESR (v_L = 110 GHz) homogeneous line width of the polymeric line on the Cs concentration in powder $Na_2Rb_{1-x}Cs_xC_{60}$ samples [93]. The homogenous line width first slightly increases on going from $x = 0$ to $x = 0.2$ and then decreases for the $x = 0.5$ and $x = 0.7$ samples, in striking contrast to the behaviour of the homogeneous line width of the cubic phase (Fig. 27 inset). The behaviour of the cubic homogeneous line width is in agreement with that reported for other A_3C_{60} systems [97] and can be understood using Eq. 2. However, in order to understand the behaviour of the polymeric phase with increasing Cs content x,

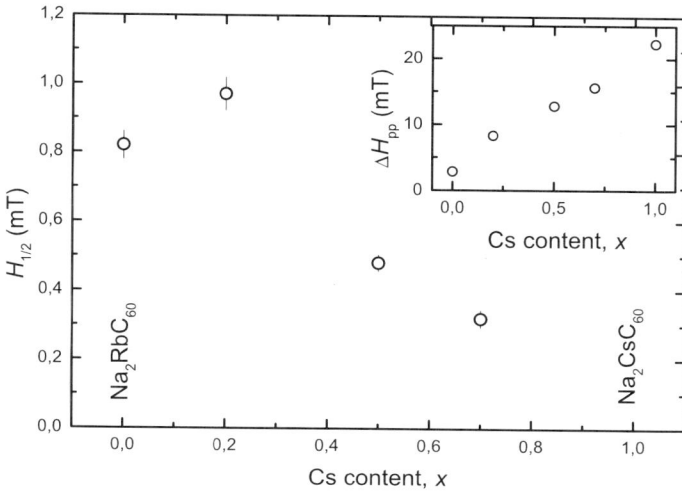

Fig. 27. Variation of the homogeneous line width of the polymeric line in $Na_2Rb_{1-x}Cs_xC_{60}$ samples as a function of the Cs content, x. The inset shows the dependence of the homogeneous ESR line width of the cubic non-polymeric phase

the variation of the parameter α with x should also be considered. Thus, the decrease in the homogeneous line width of the polymeric phase means that despite the large increase in the spin-orbit coupling constant, λ, with increasing Cs content, α decreases even more substantially. This suggests that the electronic structure of $Na_2Rb_{0.3}Cs_{0.7}C_{60}$ is much more quasi-one-dimensional than that of Na_2RbC_{60}.

We note at this point that in the analysis of the temperature dependence of the spin-lattice relaxation times in $Na_2Rb_{1-x}Cs_xC_{60}$, other relaxation channels, which may be active in these systems should also be taken into account. Namely, when x increases, the fraction of the polymeric phase decreases and consequently the coherence length within which the electron motion is restricted decreases with increasing x. Such a restricted type of electronic diffusion may additionally affect the electron spin-lattice relaxation. The second effect relates to the coupling of the spins in the polymer phase with those of the coexisting cubic phase. Determination of the individual contributions to the relaxation mechanism necessitates a systematic study of the electron spin-lattice relaxation with the pulsed ESR technique.

It was thus of no surprise to encounter the presence of a magnetic instability in $Na_2Rb_{0.3}Cs_{0.7}C_{60}$ [99]. On cooling, the ESR susceptibility, $\chi(T)$, of $Na_2Rb_{0.3}Cs_{0.7}C_{60}$ first increases slightly below 160 K, reaching a maximum between 45 and 50 K (Fig. 28). In a similar fashion to $(NH_3)K_3C_{60}$, the increase of the susceptibility between 160 K and 50 K cannot be described by the Curie-Weiss law and its origin is ascribed to antiferromagnetic correlations in a narrow band metallic system. Below this temperature, $\chi(T)$ suddenly drops and almost completely vanishes below 10 K, providing evidence of the opening

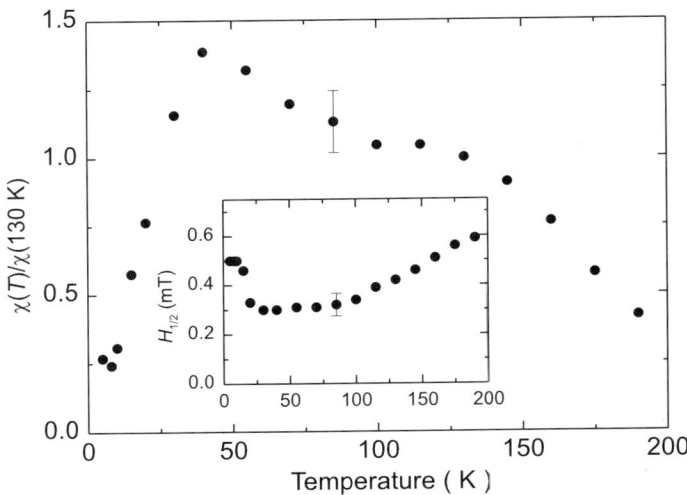

Fig. 28. Temperature dependence of the ESR susceptibility in the $Na_2Rb_{0.3}Cs_{0.7}C_{60}$ polymer phase. Inset: temperature dependence of the homogeneous width of the polymer line

of a gap at the Fermi level. The observed changes in $\chi(T)$ are also accompanied by changes in the temperature dependence of the homogeneous line width (Fig. 28 inset). $H_{1/2}$ changes considerably with temperature, decreasing almost linearly with decreasing temperature between 200 and 45 K, where it reaches a broad minimum. Below 45 K, the trend is suddenly reversed and $H_{1/2}$ starts to increase below 25 K. The antiferromagnetic fluctuations at the Larmor frequency may be responsible for the increase in the homogeneous line width at low temperatures.

There are at least three possible scenarios, which can lead to the vanishingly small ESR susceptibility observed in $Na_2Rb_{0.3}Cs_{0.7}C_{60}$ at low temperature:

(a) 3D spin density wave (SDW) ordering, as originally suggested for the RbC_{60} polymer [57],
(b) antiferromagnetic (AFM) ordering, as seen in $(NH_3)K_3C_{60}$ [49], and
(c) transition to a non-magnetic charge density wave (CDW) state, as theoretically suggested for polymeric Na_2RbC_{60} [100].

If SDW or AFM ordering occurs, the low-temperature state has a well-defined magnetic character and one expects to find an additional antiferromagnetic resonance (AFMR) mode at high frequencies. A closer inspection of the low-temperature integrated HF ESR spectra of $Na_2Rb_{0.3}Cs_{0.7}C_{60}$ indeed revealed a new line emerging on the low-field side below 15 K. A low field shift and a broadening of the new line with decreasing temperature are characteristics of both the AFMR and the resonance in spin glasses [28]. Additional frequency-dependent experiments are needed to unambiguously assign the new line to AFMR.

In this section, we have shown that the structural, electronic, and magnetic properties of the $Na_2Rb_{1-x}Cs_xC_{60}$ polymeric phases strongly depend on the Cs content. The electronic structure becomes progressively quasi-one-dimensional, as x approaches 1. Consistent with this, Na_2RbC_{60} remains conducting down to 4 K with no low-temperature ground-state instabilities, which may have led to the formation of either a SDW or a CDW state. Similar behaviour is displayed by the quaternary fulleride polymers with increasing interchain separation, despite the increased one-dimensional character, up to and including the phase with $x = 0.5$. However, in the $Na_2Rb_{0.3}Cs_{0.7}C_{60}$ salt, a pseudo-gap of magnetic origin opens at the Fermi level below 45 K. The presence of an antiferromagnetic resonance in high-field ESR below $T_N \sim 15$ K provides evidence of the onset of three-dimensional magnetic ordering. The $Na_2Rb_{1-x}Cs_xC_{60}$ family of polymers thus offers a unique way of chemical control of the electronic properties, as the opening of the gap in this system of predominantly itinerant electrons is an extremely sensitive function of the interchain separation, which is controlled by the Cs content, x.

5
Summary and Future Prospects

A common theme for all the intercalated C_{60} systems described in this chapter has been the interplay between magnetic instabilities and the metal-insulator transition and suppression of superconductivity. As the Coulomb interaction, U between two electrons on the same C_{60} molecule is by a factor of 2-3 larger than the width, W of the partially filled t_{1u} band [101] and the (U/W) ratio exceeds the critical value for which the systems should become Mott-Hubbard insulators, the fullerene intercalation compounds are highly correlated electron systems [5]. Then the electronic properties of the systems described here result from the combined effects of low-dimensional structures and electron correlation. A recent study of the phase diagram of ammoniated alkali fullerides, $(NH_3)K_{3-x}Rb_xC_{60}$ ($x = 1, 2, 3$) [102] showed that the Néel temperature, T_N, first increases with increasing interfullerene spacing and then decreases, showing a maximum at ~ 76 K for $(NH_3)KRb_2C_{60}$. The results were explained within the framework of the generalised phase diagram of the Mott-Hubbard transition with an antiferromagnetic ground state.

Work on solid fullerenes and their numerous derivatives continues unabated to produce novel and unexpected results. C_{60}-based materials display superconductivity at T_Cs still only surpassed by the high-T_C cuprate superconductors. The recent findings of a variety of magnetic phenomena in systems comprising different fullerene structural architectures has generated a surge of interest in fundamental aspects of their physics and chemistry as well as in technological applications. Although hopes for technological applications are now shifting towards carbon nanotubes [103], the simplicity of the C_{60} geometry and the variety of electronic properties found in its compounds keep this fascinating molecule in the mainstream of current research. The experience and the results obtained on C_{60}-based materials represent invaluable insight, which could be rapidly transferred to other nanostructured materials in the future.

6
References

1. See, for instance: Miller JS, Epstein AJ (1994) Angew. Chem 33: 385
2. Kroto HW, Heath JR, O'Brien SC, Curl RF, Smalley RE (1985) Nature 318: 162
3. Hayden GW, Mele EJ (1987) Phys Rev B 36: 5010; Erwin S, Pickett WE (1991) Science 254: 842
4. Hebard A (1992) Physics Today, November: 26
5. The reader is referred to the numerous reviews of superconductivity in A_3C_{60}, for instance: Gunnarsson O (1997) Rev Mod Phys 69: 575; Prassides K (1997) Current Opinion in Solid State and Materials Science 2: 433; Rosseinsky MJ (1998) Chem Mater 10: 2665; Prassides K, Margadonna S (2000) In: Kadish KM, Ruoff RS (eds) Fullerenes – Chemistry, Physics, and Technology. Wiley, New York, chap 14
6. Allemand P-M, Khemani KC, Koch A, Wudl F, Holczer K, Donovan S, Gruner G, Thompson JD (1991) Science 253: 301
7. Takahashi M, Turek P, Nakazawa Y, Tamura M, Nozawa K, Shiomi D, Ishikawa M, Kinoshita M (1991) Phys Rev Lett 67: 746
8. Tanaka K, Zakhidov AA, Yoshizawa K, Okahara K, Yamabe T, Yakushi K, Kikuchi K, Suzuki S, Ikemoto I, Achiba Y (1992) Phys Lett A164: 221; Tanaka K, Zakhidov AA, Yoshizawa K, Okahara K, Yamabe T, Yasushi K (1993) Phys Rev B 47: 7554
9. Venturini P, Mihailovic D, Blinc R, Cevc P, Dolinšek J, Abramic D, Zalar B, Oshio H, Allemand PM, Hirsch A, Wudl F (1992) J Mod Phys B 6: 3947
10. Blinc R, Pokhodnia K, Cevc P, Arcon D, Omerzu A, Mihailovic D, Venturini P, Golic L, Trontelj Z, Lužnik J, Pirnat J (1996) Phys Rev Lett 76: 523
11. Heiney PA, Fischer JE, McGhie A, Romanow W, Denenstein A, McCauley J, Smith A (1991) Phys Rev Lett 66: 2991
12. Stephens PW, Cox D, Lauher JW, Mihaly L, Wiley JB, Allemand PM, Hirsch A, Holczer K, Li Q, Thompson JD, Wudl F (1991) Nature (London) 355: 331
13. Bommeli F, Degiorgi L, Wachter P, Mihailovic D, Hassanien A, Venturini P, Schreiber M, Diedrich F (1995) Phys Rev B 51: 1366
14. Schilder A, Klos H, Rystau I, Schütz W, Gotschy B (1994) Phys Rev Lett 73: 1299
15. Omerzu A, Mihailovic D, Biskup N, Milat O, Tomic S (1996) Phys Rev Lett 77: 2045
16. Lappas A, Prassides K, Vavekis K, Arcon D, Blinc R, Cevc P, Amato A, Feyerherm R, Gygax FN, Schenck A (1995) Science 267: 1799
17. Schenck A (1993) In: Gupta LC, Multani MS (eds) Frontiers in Solid State Science. World Scientific, Singapore, vol. 2, p 269
18. Prassides K (1997) Hyperfine Interactions 106: 125
19. Arcon D, Dolinšek J, Blinc R (1996) Phys Rev B 53: 9137
20. Mihailovic D, Arcon D, Venturini P, Blinc R, Omerzu A, Cevc P (1995) Science 268: 400
21. Tanaka K, Asai Y, Sato T, Kuga T, Yamabe T, Tokumoto M (1996) Chem Phys Lett 259: 574
22. Mrzel A, Cevc P, Omerzu A, Mihailovic D (1996) Phys Rev B 53: R2922; Arcon D, Blinc R, Mihailovic D, Omerzu A, Cevc P (1999) EuroPhys Lett 46: 667
23. Arcon D, Blinc R, Cevc P, Omerzu A (1999) Phys Rev B 59: 5247
24. Omerzu A et al. (to be published)
25. David WIF, Ibberson RM, Matthewman JC, Prassides K, Dennis TJS, Hare JP, Kroto HW, Taylor JP, Walton DRM (1991) Nature 353: 147; David WIF, Ibberson RM, Dennis TJS, Hare JP, Prassides K (1992) EuroPhys Lett 18: 219
26. Kambe T, Nogami Y, Oshima K (2000) Phys Rev B 61: R862
27. Vonsovskii SV (1996) Ferromagnetic Resonance. Pergamon Press, Oxford; Arcon D, Blinc R, Omerzu A (1997) Molecular Physics Reports 18/19: 89
28. Halperin BI, Saslow WM (1977) Phys Rev B 16: 2154; Saslow WM (1980) Phys Rev B 22: 1174; Saslow WM (1982) Phys Rev Lett 48: 505
29. Gatteschi D, Caneschi A, Pardi L, Sessoli R (1994) Science 265: 1054
30. Arcon D, Cevc P, Omerzu A, Blinc R (1998) Phys Rev Lett 80: 1529

31. Blinc R, Cevc P, Arcon D, Omerzu A, Mehring M, Knorr S, Grupp A, Barra A-L, Chouteau G (1998) Phys Rev B 58: 14416
32. Tanigaki K, Ebbsen TW, Saito S, Mizuki J, Tsai JS, Kubo Y, Kuroshima S (1991) Nature 352: 222
33. Fleming RM, Ramirez AP, Rosseinsky MJ, Murphy DW, Haddon RC, Zahurak SM, Makhija AV (1991) Nature 352: 787
34. Zhou O, Fleming RM, Murphy DW, Rosseinsky MJ, Ramirez AP, van Dover RB, Haddon RC (1993) Nature 362: 433
35. Rosseinsky MJ, Murphy DW, Fleming RM, Zhou O (1993) Nature 364: 425
36. Zhou O, Palstra TTM, Iwasa Y, Fleming RM, Hebard AF, Sulewski PE (1995) Phys Rev B 52: 483
37. Ishii K, Watanuki T, Fujiwara A, Suematsu H, Iwasa Y, Shimoda H, Mitani T, Nakao H, Fujii Y, Murakami Y, Kawada H (1999) Phys Rev B 59: 3956
38. Brown CM, Prassides K, Iwasa Y, and Shimoda H (1997) In: Kadish KM, Ruoff RS (eds) Recent Advances in the Chemistry and Physics of Fullerenes and Related Materials. Electrochemical Society, Pennington, vol 4, p 1224
39. Margadonna S (2000) D.Phil Thesis, University of Sussex
40. Iwasa Y, Shimoda H, Palstra TTM, Maniwa Y, Zhou O, Mitani T (1996) Phys Rev B 53: R8836
41. Allen KM, Heyes SJ, Rosseinsky MJ (1996) J Mater Chem 6: 1445
42. Margadonna S, Prassides K, Neumann DA, Shimoda H, Iwasa Y (1999) Phys Rev B 59: 943
43. Ricco M and Arcon D (unpublished data)
44. Simon F, Janossy A, Iwasa Y, Shimoda H, Baumgartner G, Forro L (1998) In: Kuzmany H, Fink J, Mehring M, Roth S (eds) Electronic Properties of Novel Materials. AIP Conference Proceedings, Woodbury, NY, vol 442, p 296
45. Ramirez AP et al. (1992) Phys Rev Lett 69: 1687
46. Petit P, Robert J (1996) Appl Magn Reson 11: 183
47. Simon F, Janossy A, Muranyi F, Feher T, Shimoda H, Iwasa Y, Forro L (2000) Phys Rev B 61: R3826
48. Prassides K, Tanigaki K, Iwasa Y (1997) Physica C 282: 307
49. Prassides K, Margadonna S, Arcon D, Lappas A, Shimoda H, Iwasa Y (1999) J Am Chem Soc 121: 11227
50. Tou H, Maniwa Y, Iwasa Y, Shimoda H, Mitani T (2000) Phys Rev B 62: R775
51. Koch E, Gunnarsson O, Martin RM (1999) Phys Rev Lett 83: 620
52. Murphy DW et al. (1992) J Phys Chem Solids 53: 1321
53. Winter J, Kuzmany H (1992) Solid State Commun 84: 935
54. Poirier DM, Weaver JH (1993) Phys Rev B 47: 10959
55. Zhu Q et al. (1993) Phys Rev B 47: 13948
56. Janossy A et al. (1993) Phys Rev Lett 71: 1091
57. Chauvet O, Oszlanyi G, Forro L, Stephens PW, Tegze M, Faigel G, Janossy A (1994) Phys Rev Lett 72: 2721
58. Stephens PW et al. (1994) Nature 370: 636
59. Launois P, Moret R, Hone J, Zettl A (1998) Phys Rev Lett 81: 4420
60. Rouzière S, Margadonna S, Prassides K, Fitch AN (2000) EuroPhys Lett 51: 314
61. (a) Zhu Q, Cox DE, Fischer JE (1995) Phys Rev B 51: 3966; (b) Oszlanyi G, Bortel G, Faigel G, Tegze M, Granasy L, Pekker S, Stephens PW, Bendele G, Dinnebier R, Mihaly G, Janossy A, Chauvet O, Forro L (1995) Phys Rev B 51: 12228; (c) Oszlanyi G, Bortel G, Faigel G, Granasy L, Bendele G, Stephens PW, Forro L (1996) Phys Rev B 54: 11849
62. Kosaka M, Tanigaki K, Tanaka T, Take T, Lappas A, Prassides K (1995) Phys Rev B 51: 12018
63. Lappas A, Kosaka M, Tanigaki K, Prassides K (1995) J Am Chem Soc 117: 7560
64. For a recent review see: Bennati M, Griffin RG, Knorr S, Grupp A, Mehring M (1998) Phys Rev B 58: 15603
65. Bommeli F et al. (1995) Phys Rev B 51: 14794
66. Pekker S et al. (1994) Solid State Commun 90: 349

67. Brouet V et al. (1996) Phys Rev Lett 76: 3638
68. Jannosy A, Nemes N, Feher T, Oszlanyi G, Baumgartner G, Forro L (1997) Phys Rev Lett 79: 2718
69. Erwin SC, Krishna GV, Mele EJ (1995) Phys Rev B 51: 7345
70. Auban-Senzier P et al. (1996) J Phys I 6: 2181
71. Tycko R, Dabbagh G, Murphy DW, Zhu Q, Fischer JE (1993) Phys Rev B 48: 9097
72. McFarlane WA, Kiefl RF, Dunsiger S, Sonier JE, Fischer JE (1995) Phys Rev B 52: R6995
73. Uemura YJ, Kojima K, Luke GM, Wu WD, Oszlanyi G, Chauvet O, Forro L (1995) Phys Rev B 52: R6991
74. Cristofolini L, Lappas A, Vavekis K, Prassides K, DeRenzi R, Ricco M, Schenck A, Amato A, Gygax FN, Kosaka M, Tanigaki K (1995) J Phys Condens Matter 7: L567
75. Simovic B, Jerome D, Rachdi F, Baumgartner G, Forro L (1999) Phys Rev Lett 82: 2298
76. Bourbonnais C, Dumoulin B (1996) J Phys I France 6: 1727
77. Doumlin B, Bourbonnais C, Ravy S, Pouget JP, Coulon C (1996) Phys Rev Lett 76: 1360
78. Dumm M, Loidl A, Fravel BW, Starkey KP, Montgomery LK, Dressel M (2000) Phys Rev B 61: 511
79. Pouget JP, Ravy S (1997) Synth Metals 85: 1523
80. Brouet V, Alloul H, Quere F, Baumgartner G, Forro L (1999) Phys Rev Lett 82: 2131
81. Stephens PW, Mihaly L, Lee PL, Whetten RL, Huang SM, Kaner R, Deiderich F, Holczer K (1991) Nature 351: 632
82. Kniaz K, Fischer JE, Zhu Q, Rossiensky MJ, Zhou O, Murphy DW (1993) Solid State Commun 88: 47
83. Prassides K, Christides C, Thomas IM, Mizuki J, Tanigaki K, Hirosawa I, Ebbesen TW (1994) Science 263: 950
84. Mizuki J, Takai M, Takahashi H, Môri N, Hirosawa I, Tanigaki K, Prassides K (1994) Phys Rev B 50: 3466.
85. Yildirim T, Fischer JE, Dinnebier R, Stephens PW, Lin CL (1995) Solid State Commun 93: 269
86. Prassides K, Vavekis K, Kordatos K, Tanigaki K, Bendele GM, Stephens PW (1997) J Am Chem Soc 119: 834
87. Bendele GM, Stephens PW, Prassides K, Vavekis K, Kordatos K, Tanigaki K (1998) Phys Rev Lett 80: 736
88. Lappas A, Brown CM, Kordatos K, Suard E, Tanigaki K, Prassides K (1999) J Phys Condens Matter 11: 371
89. Margadonna S, Brown CM, Lappas A, Prassides K, Tanigaki K, Knudsen KD, LeBihan T, Mézouar M (1999) J Solid State Chem 145: 471
90. Brown CM, Takenobu T, Kordatos K, Prassides K, Iwasa Y, Tanigaki K (1999) Phys Rev B 59: 4439
91. Prassides K, Brown CM, Margadonna S, Kordatos K, Tanigaki K, Suard E, Dianoux AJ, Knudsen KD (2000) J Mater Chem 10: 1443
92. Arcon D, Prassides K, Margadonna S, Maniero AL, Brunel LC, Tanigaki K (1999) Phys Rev B 60: 3856
93. Arcon D, Prassides K, Maniero AL, Brunel LC Appl Magn Res (in press)
94. Simon F et al. (unpublished)
95. Gatteschi D, Sessoli R (1990) Magn Res Rev 15: 1
96. Tanigaki K, Kosaka M, Manako T, Kubo Y, Hirosawa I, Uchida K, Prassides K (1995) Chem Phys Lett 240: 627
97. Elliot RJ (1954) Phys Rev 96: 266
98. Wegner et al. (1978) J Phys C 39: 1454
99. Arcon D, Prassides K, Maniero AL, Brunel LC (2000) Phys Rev Lett 84: 562
100. Surjan PR, Lazar A, Kallay M (1998) Phys Rev B 58: 3490
101. Gunnarsson O, Koch E, Martin M (1996) Phys Rev B 54: R11026
102. Takenobu T, Muro T, Iwasa Y, Mitani T (2000) Phys Rev Lett 85: 381
103. Narymbetov B, Omerzu A, Kabanov VK, Tokumoto M, Kobayashi H, Mihailovic D (2000) Nature 407, p 883

Molecular Compounds Showing a Spin Ladder Behaviour

Concepció Rovira

Institut de Ciència de Materials de Barcelona (CSIC), Campus Universitari, 08193, Bellaterra, Spain
E-mail: cun@icmab.es

Spin ladders are low-dimensional magnetic quantum systems that were recently discovered. A comprehensive overview of their structural and magnetic characteristics is summarised here and the possibilities for supramolecular chemistry in the development of such fascinating materials are discussed. By using molecules as spin carrier units the interacting spins are located in molecular orbitals with a dominant role of π electrons. On the basis of a review of the few existing examples of molecular spin ladders it is clear that typical supramolecular and crystal engineering criteria, such as π-π overlap, S$\cdots$S and C—H$\cdots$S interactions and the complementary nature of size and shape, are useful in the construction of these magnetic quantum systems with intermediate dimensionalities between 1D chains and 2D square lattices.

Keywords: Spin ladders, Molecular magnetism, Crystal engineering, Supramolecular chemistry, Radical ions, Molecular materials

1	Introduction	163
2	Characteristics of Spin Ladders	164
3	Molecular Spin Ladders	168
3.1	Molecular Spin Ladders by the Assembling of Chains	169
3.2	Molecular Spin Ladders by the Assembling of Dimers	179
4	Perspectives	186
5	References	187

1
Introduction

Even though the field of magnetic materials has been traditionally confined to metals, molecule-based magnets recently appeared on the scene of magnetism. The rational design of molecular magnetic materials has permitted the modulation of their structural and magnetic dimensionalities and, on going from 1D to 2D and finally 3D spin-coupled systems, the stabilisation of bulk ferromagnetism in many of these compounds has been achieved [1].

The main interest of these new materials arises from the fact that the interacting spins are located in molecular orbitals (MO) with a large or dominant role of s and p atomic orbitals rather than of d and f orbitals as in ordinary metals and magnets. Moreover, the modulation of their crystal structures and thereby of their electronic and magnetic properties and dimensionalities can be achieved by modifying the type and/or the strength of intermolecular interactions present in the crystal. By changing the structure and components of the molecular units, modifications of this type can, in principle, be accomplished although the methodologies required for the control of the crystal packing of molecular materials as well as the factors that govern the magnetic interactions between purely organic molecules are still not completely understood. Nevertheless, the flurry of activity in this field can be traced to the widely held belief that even the most sophisticated properties, and combinations of them, can be rationally designed from the bottom up. This motivation was further fuelled by the increased synthetic capabilities that permit us to synthesise molecules with suitable structures and topologies, and also by the spectacular progress of supramolecular chemistry for materials development as witnessed in the recent years. As a result of a systematic synthetic effort, it has been possible to obtain different kinds of crystalline supramolecular organisations, using open-shell molecular building blocks, that show a richly diverse array of magnetic properties such as antiferromagnetism, ferrimagnetism, kagome antiferromagnetism, weak ferromagnetism, 2D and 3D ferromagnetism, and spin-ladder behaviour. The combination of magnetic properties with others, such as electrical or optical properties, has also been achieved [1].

This chapter is devoted to the very few molecular materials showing spin-ladder behaviour, describing first the structural and magnetic characteristics necessary to achieve this recently discovered, fascinating physical property.

2
Characteristics of Spin Ladders

Spin ladders are low-dimensional magnetic quantum systems that consist of a finite number of strongly magnetically coupled chains of spins that are at the crossroad between one (1D) and two dimensions (2D) (Fig. 1). In principle, one might expect that a smooth crossover from chains to a plane would result if one assembled chains to form ladders of increasing width. But, in fact, the field of ladder systems emerged strongly in magnetism when Dagotto et al. [2] found, by numerical calculations, that the crossover from the quasi-long-range order in a chain (1D) of antiferromagnetically coupled $S = 1/2$ spins to the true long-range order that occurs in a plain (2D) is not at all smooth.

The ladders with $S = 1/2$ spins that show antiferromagnetic isotropic coupling between the nearest-neighbours, and that are generally named as $S = 1/2$ Heisenberg antiferromagnetic spin ladders are especially interesting. The Hamiltonian in Eq. (1), defines such magnetic systems,

$$H = -J_\parallel \sum_{a=1,2} \sum_{i=1}^{L} S_{i,a} \cdot S_{i+1,a} - J_\perp \sum_{i=1}^{L} S_{i,1} \cdot S_{i,2} \qquad (1)$$

Molecular Compounds Showing a Spin Ladder Behaviour

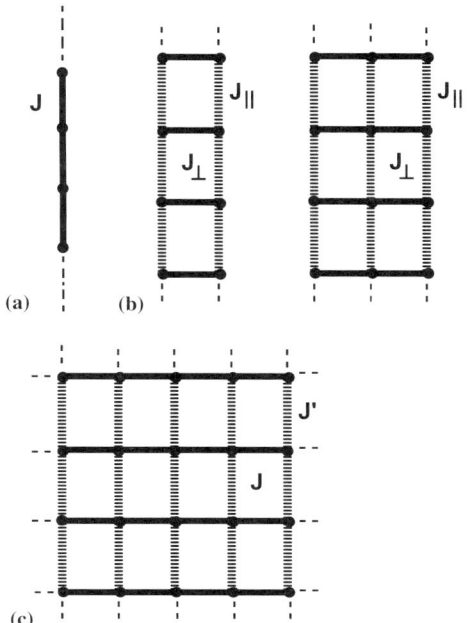

Fig. 1. Schematic illustration of different kinds of low dimensional magnetic Heisenberg systems with antiferromagnetic couplings: (**a**) 1D chain; (**b**) two-leg and three-leg spin ladders where $J_\perp$ is the coupling along the rungs and $J_\parallel$ is the coupling along the chains; (**c**) 2D square lattice with different exchange coupling constants, J and J'. Black dots denote $S = 1/2$ spin-containing units and continuous and dashed lines represent antiferromagnetic interactions of different strength

where $S_{i,a}$ is the spin operator at the site i ($i = 1, 2, \ldots, L$) on the leg a ($a = 1, 2 \ldots$) of a ladder with L rungs. $J_\perp$ and $J_\parallel$ denote the intra- and inter-rung exchange couplings, respectively. With this definition, $J_\perp$ and $J_\parallel$ should be negative for antiferromagnetic interactions.

The magnetic properties of such systems are particularly interesting as a result of the increased importance of the quantum mechanical effects operating within them. Theoretical calculations have predicted that spin ladders with an even number of legs have a spin-liquid ground state, so called because of their purely short-range spin correlation along the legs. These even-leg ladders consist of spin singlet pairs with a spin-spin correlation distance along the legs that show an exponential decay produced by the presence of a finite spin gap. By contrast, a ladder with an odd number of legs behaves quite differently and displays properties similar to those of a 1D antiferromagnetic Heisenberg chain at a low thermal energy, namely, gapless spin excitations and a power-law falloff of the spin-spin correlations which are magnetically ordered [3–10].

The magnetic properties of two-leg $S = 1/2$ antiferromagnetic spin ladders can easily be described in the simple limit where intra-rung exchange coupling

($J_\perp$) is much larger than the exchange coupling along the legs ($J_\parallel$); $J_\perp \gg J_\parallel$. The energy of the ground state, E_{gs}, in this limit is approximately $E_{gs} = -3/4\, J_\perp N$, where N is the number of rungs and $3/4 J_\perp$ is the energy of each rung singlet state. In this particular case, the rungs of the ladder interact only weakly with each other and the dominant spin configuration in the ground state is that with the spins on each rung forming a spin singlet. Therefore, the ground state has a total spin $S = 0$. To produce the lowest excitation in the infinite ladder to a configuration with a total spin $S = 1$, one of the rung singlets of the ladder must be promoted to an $S = 1$ triplet (Fig. 2) with a finite energy that must overcome the finite spin gap of the system. Moreover, the coupling along the legs creates an energy band of $S = 1$ magnons with a dispersion of about $2J_\parallel$ leading to an exponential decrease in the magnetic spin susceptibility as the temperature is lowered until kT is below the spin gap energy (Fig. 3).

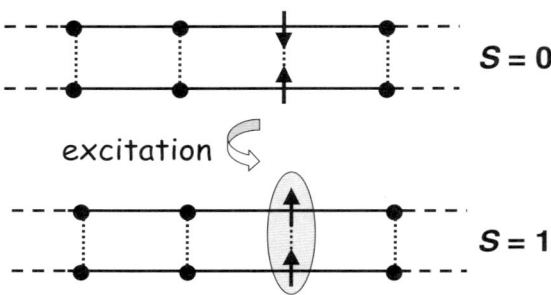

Fig. 2. Schematic illustration of the lowest excitation to a configuration with a total spin $S = 1$ in a two-leg $S = 1/2$ Heisenberg antiferromagnetic ladder with $J_\perp > J_\parallel$

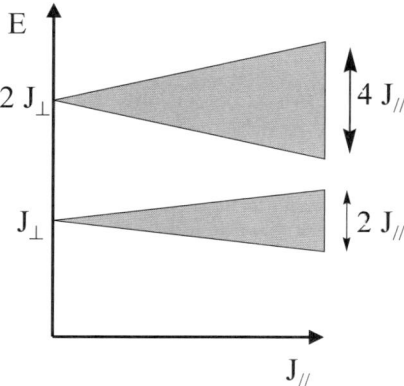

Fig. 3. Evolution of the energy levels in a spin ladder when the interaction along the legs is turned on

Another interesting prediction for two-leg ladders is that light hole doping can even lead to high Tc superconductivity on basis that effective attraction between extra holes may arise from the magnetic interactions between the spins of the ladders [2]. Odd-leg ladders, on the other hand, must have different behaviour because singlet pair formation on their rungs is not possible [5]. In fact, they display properties similar to those of single chains at low energies.

"Ideal" spin ladders are considered to be those in which the exchange coupling along the rungs is very similar to the exchange coupling along the chains, that is $J_\perp/J_\parallel \approx 1$. Therefore, regarding the two-leg spin ladders, such systems are intermediate between two limiting cases:

a) isolated dimers, when $J_\perp/J_\parallel \to \infty$, and
b) isolated 1D chains, when $J_\perp/J_\parallel \to 0$.

Although these limiting cases have been analysed from the theoretical point of view, spin ladders of experimental interest are those having both exchange coupling constants of the same order of magnitude, that is $0.1 \leq J_\perp/J_\parallel \leq 10$. Another factor that should be taken into account on considering a system as a spin ladder is that ladders should be quite well isolated one from another, since appreciable interladder coupling (J' in Fig. 4) can promote a quantum phase transition between the spin liquid ground state and a magnetically ordered state. Troyer et al., established a critical ratio of the interladder to intraladder coupling of $J'/J_\perp \approx 0.1$ that separates the spin liquid from an antiferromagnetic ordered state [11].

The appealing properties theoretically predicted for spin ladders have only recently been found in very few materials. The first examples were connected chains of transition metals in inorganic oxides, such as the series of even- and odd-leg ladder structures provided by the cuprates $Sr_{n-1}Cu_{n+1}O_{2n}$ ($n = 3, 5$) [12] or the two-leg ladders $(VO)_2P_2O_7$ [13] and $LaCuO_{2.5}$ [14]. Another system that has been described with various models including a spin ladder one is $Cu_2(C_5H_2N_2)_2Cl_4$ with interacting Cu atoms being in the strong coupling limit ($J_\perp/J_\parallel = 5.5$) [9]. However, it is important to note that the identification of

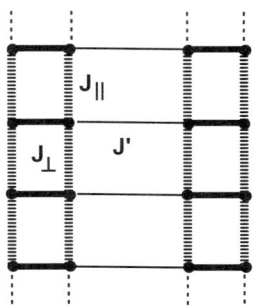

Fig. 4. Scheme of two-leg ladders showing intra-ladder ($J_\perp$ and $J_\parallel$) and inter-ladder (J') exchange couplings. Black dots denote $S = 1/2$ spin-containing units

these compounds as spin ladders has been made, primarily, on the basis of their macroscopic characterisation (magnetic susceptibility data) and it has been shown later on that microscopic probes, such as neutron scattering, nuclear magnetic resonance (NMR), and muon spin resonance (μSR), are necessary to establish the exact nature of the ground state. In fact, although the family of cuprates containing Sr has been confirmed as real spin ladders by different experiments [15], NMR and μSR studies on the La-containing cuprate[16, 17] and inelastic neutron scattering measurements on the $(VO)_2P_2O_7$ [18] showed that both compounds are not real spin ladders.

The search for new compounds showing spin ladder properties is a very important task not only for understanding the fascinating quantum ladder physics but also for developing new materials, especially superconductors. Recently, some cuprates that behave as spin ladders have shown superconducting properties upon doping [19].

3
Molecular Spin Ladders

In the search for new spin ladder compounds, those belonging to the family of molecular solids can be of great interest since they have a new molecular architecture for interacting spins. In fact, in the field of molecular materials science and from the viewpoint of chemists, one of the most exciting topics is solid state supramolecular synthesis; that is to combine appropriate molecular building blocks in a predesigned way so as to endow the resulting supramolecular assembly with desirable physical properties. Because of their structural flexibility, molecular solids offer a field of choice to finely tune interesting physical properties such as those of the spin ladder materials. It should be emphasised that, in materials constructed from open-shell molecules, the modulation of their structural, electronic, and magnetic dimensionalities has permitted the realisation of molecular compounds showing a large variety of physical phenomena ranging from low-dimensional metals and superconductors to organic ferromagnets [1, 20]. As the interacting spins in these materials are located in molecular orbitals (MO), the molecule-based magnetic systems should be commonly understood by using the MO concepts and to analyse the magnetic interactions it is not only the connectivity between atoms that we should be considering, but also the connectivity between orbitals. Thus, the two most important issues to be considered for the creation of molecular spin ladders are:

a) the election of the spin-containing building blocks, or open-shell repeating units (molecules), and
b) the assembling of such molecular building blocks or repeating units leading to correct connectivity of orbitals.

The construction of the simplest molecular spin ladder – a two-leg ladder – can be conceptually achieved following two different strategies (Fig. 5). The first one consists of the connection of two molecular 1D $S = 1/2$ spin chains, one next to the other, and the second one to assemble an infinite number of

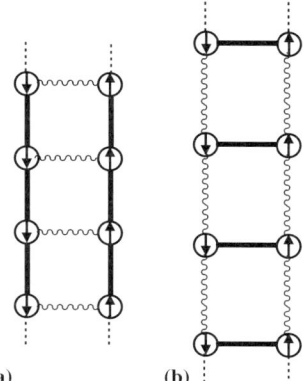

Fig. 5. Illustration of the two strategies that can be used in the construction of molecular two-leg ladders: (a) assembling of 1D $S = 1/2$ chains, and (b) connecting of $S = 1/2$ dimers

units with two interacting $S = 1/2$ molecules (dimers). Consequently, the election of the appropriate open-shell building blocks, able to form 1D chains or dimers and that are also able to interact with the neighbouring units giving rise to antiferromagnetic interactions, is of primary importance. Although the few molecular spin-ladders known so far have been obtained by serendipity, the study of these compounds shows that both conceptual strategies have been involved in their construction [21–24]. In the next sections the very few already known molecular spin ladders constructed following the two mentioned strategies will be described.

3.1
Molecular Spin Ladders by the Assembling of Chains

To follow the first strategy – the assembly of chains – one of the most promising building blocks are tetrathiafulvalene (TTF) derivatives (Scheme 1) since it is well known that these planar π-electron donors form stable radical cations, by donating an electron from the HOMO, and have a large tendency to stack, thus forming 1D chains [20].

This packing promotes the overlap of directional π-orbitals along the stacking direction that results in anisotropic band structures. Furthermore, the transfer integrals (t) between neighbouring molecules, which describe the intermolecular electronic interactions, are comparable or even smaller than

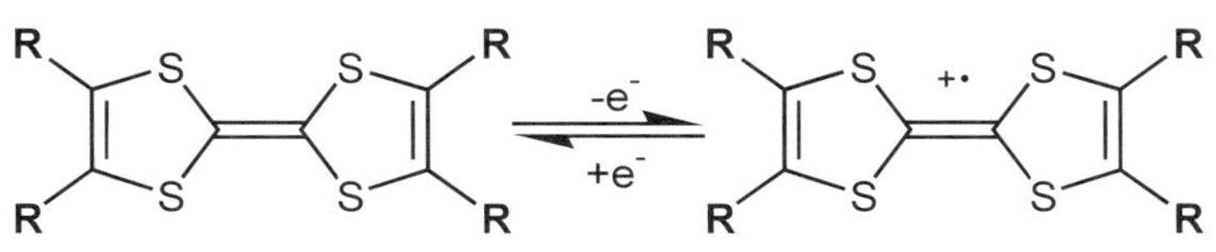

Scheme 1.

the on-site Coulombic repulsion (U) and thus the electron wave function is very often on the borderline between the localised and delocalised regimes. When localisation of the electrons is promoted, by lowering the t/U ratio, a 1D spin chain can be generated that could provide the legs of the ladders if interchain magnetic interactions are adequate. In fact, the sulfur atoms of TTF derivatives, that have a large contribution to the HOMO of the molecule, readily establish intercolumn structural and electronic interactions, thus promoting enhanced electronic dimensionality in a great number of metallic charge-transfer complexes and radical-cation salts derived from these mainly planar molecules [20, 25]. Therefore, S···S interactions could provide the rungs of the ladder by connecting the 1D chains. Isotropic exchange coupling constants, J_i, among open-shell molecules in these TTF-based molecular compounds are related with the transfer integral, t_i, of the corresponding intermolecular contact and the on-site Coulombic repulsion, U, of the molecule by Eq. [2] having predominantly an antiferromagnetic character:

$$J_i \cong 2t_i^2/U \tag{2}$$

Due to the above-mentioned characteristics, structural and electronic 1D and 2D layered, TTF-based organic conductors are quite common in the literature [20], although the achievement of solids with an intermediate structural dimensionality – that is a finite number of assembled stacks forming ladder-like structures – has only recently been accomplished [21, 22]. Only when the proper array of diamagnetic counterions cut a 2D layered TTF structure with localised electrons, will a spin ladder molecular material result in the donor sublattice. The two existing examples that will be described in the following are two-leg spin ladders obtained in such a way since the 2D donor layers are cut every two chains by diamagnetic counterions.

As a first example, the purely organic molecular two-leg spin ladder compound prepared in our group by using as a building block the TTF derivative dithiophenetetrathiafulvalene (DT-TTF) and the Au complex of the maleonitrile dithiolate [Au(mnt)$_2$]$^-$ as a magnetically innocent counterion will be described and analysed (Scheme 2) [21]. This planar diamagnetic counterion has the same size as the donor and also exhibits a great tendency to stack, so forming 1D chains, and therefore being complementary to DT-TTF.

DT-TTF and [Au(mnt)$_2$]$^-$ molecules crystallise together to form the mixed valence salt (DT-TTF)$_2$Au(mnt)$_2$ in which segregated DT-TTF (donor, D) and [Au(mnt)$_2$] (acceptor, A) stacks with a herringbone pattern are present (Fig. 6).

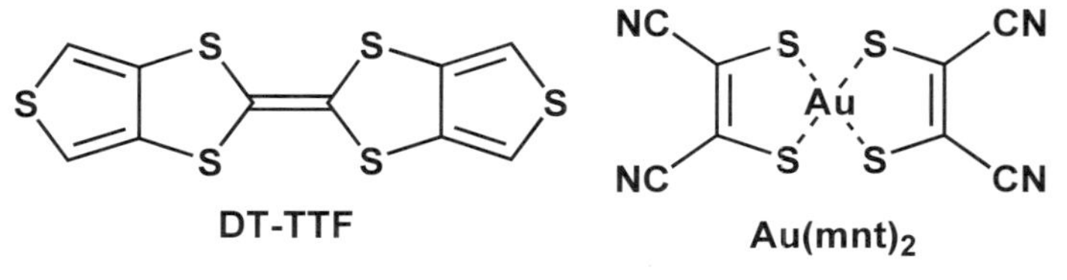

Scheme 2.

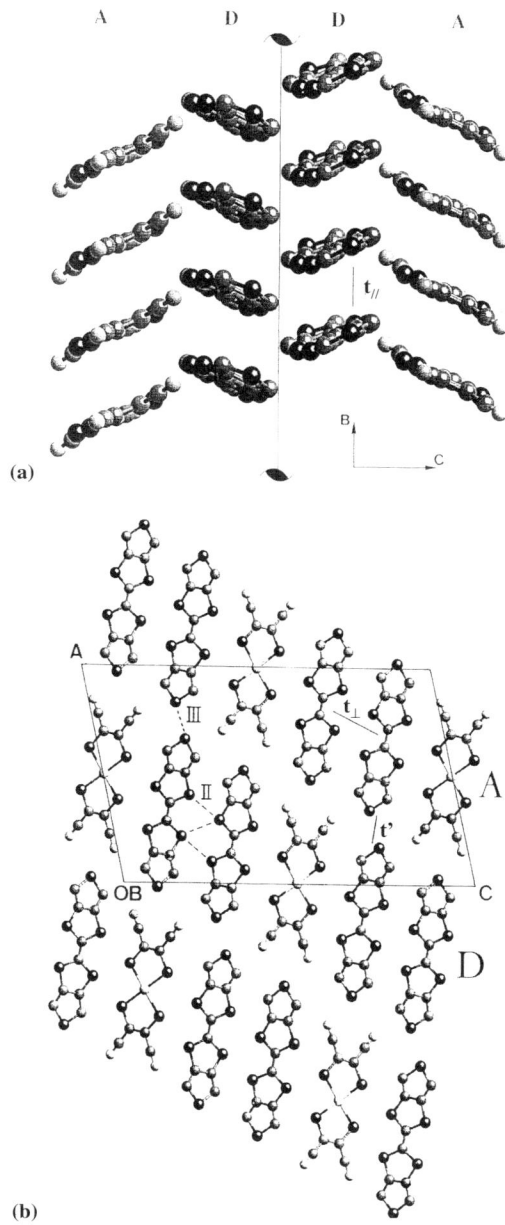

Fig. 6. (a) Crystal packing of $(DT\text{-}TTF)_2Au(mnt)_2$ showing the ladder structure formed by the organic DT-TTF donor stacks related by a two-fold screw axis; (b) projection of the crystal structure along the b axis, where short S···S contacts are indicated by dotted lines

At room temperature, the DT-TTF stacks are arranged in pairs related by a twofold screw axis, that alternate with single stacks of [Au(mnt)$_2$] units along the a-c direction. The pairs of organic donor stacks are the legs of a two-leg ladder structure since they are strongly linked by three inter-stack S···S close contacts (interaction II in Fig. 6b), providing the rungs for the ladder. In fact, the calculated transfer integrals show that the two legs of the ladder are interacting strongly since the value of the transfer integral coupling the two paired chains of donors ($t_\perp$ = 21 meV) is very similar to that of the transfer integral up the chain ($t_\parallel$ = 36 meV), the on-site Coulombic repulsions of DT-TTF being of the order of 1 eV.

The structural arrangement of donors in [(DT-TTF)$_2$Au(mnt)$_2$] seems to be governed by strong intermolecular interactions since it is the same as that found in the 2D organisation of the neutral donor in the solid state (Fig. 7) [25, 26]. Therefore, the structure of the [(DT-TTF)$_2$Au(mnt)$_2$] salt can be described by separating, in pairs of stacks, the 2D arrangement of the neutral DT-TTF with the stacks of anions. Thus, by using a diamagnetic anion of similar size and shape to that of the donor molecules, we have succeeded in the realisation of an intermediate dimensionality between 1D and 2D for a molecular compound.

This structural ladder should have localised interacting electrons in order to behave as a two-leg spin ladder; otherwise it could behave as a 1D metal. Indeed, electrical transport measurements (Fig. 8) show that below 220 K the electrons are localised and, X-ray diffuse scattering experiments (Fig. 9) confirm that a dimerisation is achieved below 220 K in the chain direction b.

In addition, the finite width of the observed X-ray diffuse lines shows that the dimerisation locally breaks the twofold screw axis symmetry that relates the two DT-TTF chains that form the ladder. In this picture, each dimer of DT-TTF molecules has one localised electron (Fig. 10). The dimers [(DT-TTF)$_2^{+\bullet}$] constitute the spin carrying units of the spin ladder.

Susceptibility data on powdered and single crystal samples of the salt [(DT-TTF)$_2$Au(mnt)$_2$] show a characteristic behaviour of localised spins with strong AF interactions, which displays a spin gap (Fig. 11). The exponential decay of susceptibility data at low temperature (from 8 to 45 K) was successfully fitted with the Troyer expression [Eq. (3)] [8] for a two-leg spin ladder system with the energy gap, Δ/k, in the spin excitation spectrum found to be equal to 78 K.

$$\chi_{\text{ladder}} = \alpha T^{-1/2} \exp(-\Delta/kT) \tag{3}$$

The exchange interactions along the legs ($J_\parallel/k$ = -83 K) and the rungs ($J_\perp/k = -142$ K) of the ladder were extracted by fitting of the whole susceptibility data (Fig. 11) with Eq. (4), which takes into account the Curie contribution due to paramagnetic impurities (defects) in the crystals and with Eq. (5). The two-leg ladder model [Eq. (5)] used for such a fitting was that developed by Barnes and Riera, where c_1, c_2, c_3, c_4, c_5, and c_6, depend on $J_\perp$ and $J_\parallel$ values [13b].

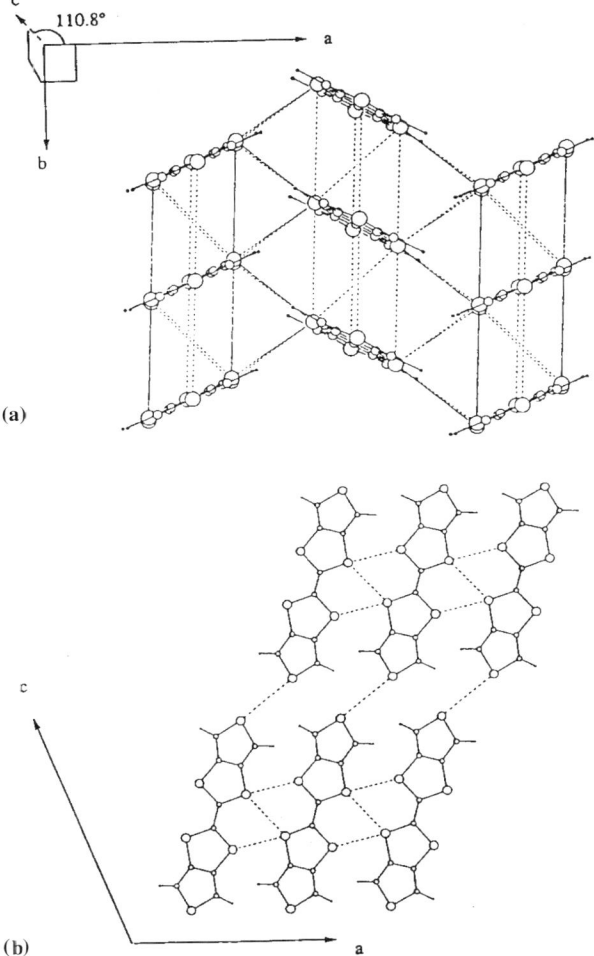

Fig. 7. Crystal packing of DT-TTF: (a) view along the long axis of the molecules showing the organic DT-TTF stacks related by a two-fold screw axis; (b) projection of the crystal structure along the b axis. Short S$\cdots$S contacts are indicated by dotted lines

$$\chi = f\chi_{\text{ladder}} + (1-f)\chi_{\text{Curie}} \qquad (4)$$

$$\chi_{\text{ladder}}(T) \frac{c_1}{T}\left[1 + \left(\frac{T}{c_2}\right)^{c_3}\left(e^{c_4/T} - 1\right)\right]^{-1}\left[1 + \left(\frac{c_5}{T}\right)^{c_6}\right]^{-1} \qquad (5)$$

A remarkable feature of this two-leg spin ladder is the fact that the $J_\perp/J_\parallel$ ratio is 1.7, which is close to that of an "ideal" spin ladder. The spin gap has also been calculated from the resulting values of $J_\parallel$ and $J_\perp$ with the theoretical

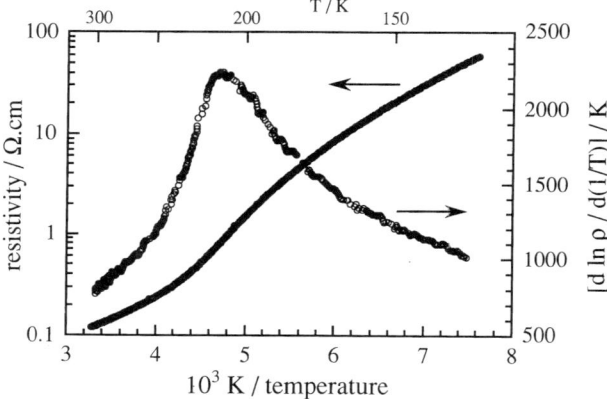

Fig. 8. Electrical resistivity ρ of [(DT-TTF)$_2$Au(mnt)$_2$] as a function of the reciprocal temperature (*left*) and its first derivative (*right*) measured along the needle axis b by the standard four-probe method

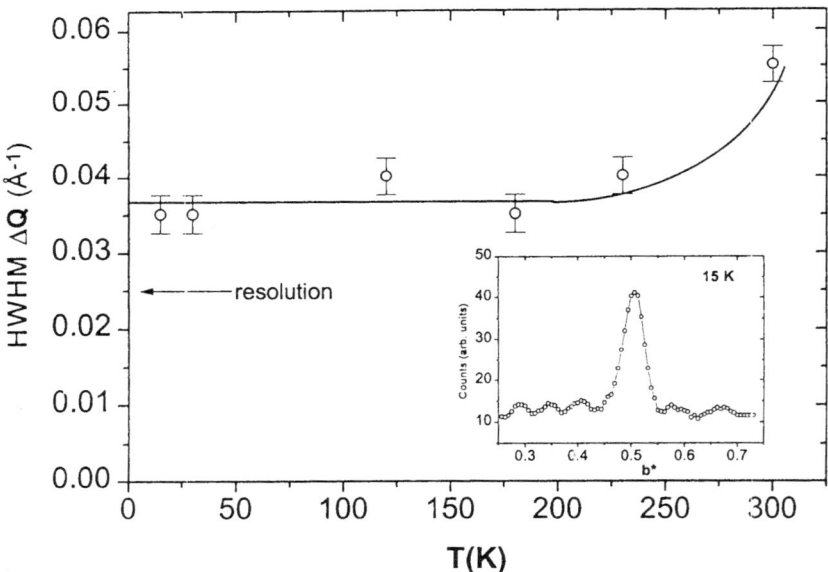

Fig. 9. Thermal dependence of the HWHM along the b axis of the 1/2b^* diffuse lines of [(DT-TTF)$_2$Au(mnt)$_2$]. The inset shows the profile along b^* of such a line (χ is the number of counts in arbitrary units)

Eq. (6) [8], giving $\Delta/k = 83$ K, which is in good agreement with the previous value.

$$\Delta = |J_\perp| - |J_\parallel| + J_\parallel^2/2J_\perp \tag{6}$$

Molecular Compounds Showing a Spin Ladder Behaviour

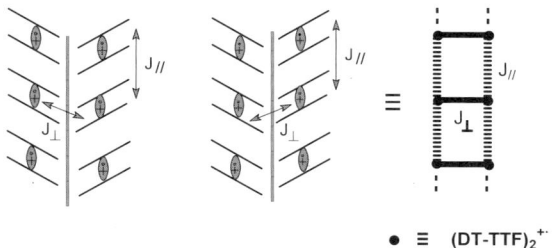

Fig. 10. Schematic illustration of the two possible two-leg ladders formed by the dimerisation of DT-TTF stacks in [(DT-TTF)$_2$Au(mnt)$_2$]

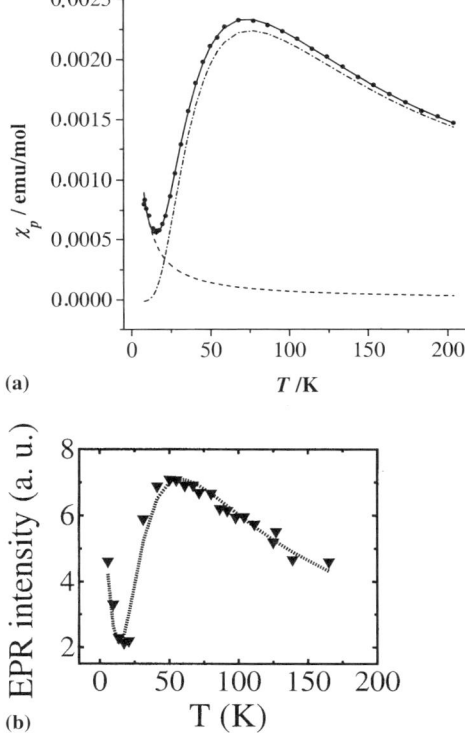

Fig. 11. (a) Temperature dependence of the paramagnetic susceptibility (●) of the (DT-TTF)$_2$Au(mnt)$_2$ salt. The solid line is the fit by Eq. (2) to a contribution of a spin ladder model (-●-●-●) and a Curie tail (- - - -). (b) Temperature dependence of the paramagnetic susceptibility (▲) of the (DT-TTF)$_2$Au(mnt)$_2$ salt measured by ESR on a single crystal

The EPR signal of a single crystal of [(DT-TTF)$_2$][Au(mnt)$_2$] has the typical parameters (g value and ΔH_{pp}) of TTF derivatives. The intensity of this EPR signal, which is proportional to the paramagnetic spin susceptibility, also exhibits the same thermal dependence as the static spin susceptibility

(Fig. 11b), confirming that the spin-ladder behaviour occurs on the organic [(DT-TTF)$_2^{-\bullet}$] stacks.

The temperature dependence of the magnetic susceptibility clearly inferred that the [(DT-TTF)$_2$Au(mnt)$_2$] salt is a purely organic molecular system with a two-leg spin ladder configuration. Nevertheless, since the calculated inter-ladder transfer integral $t' = 6$ meV, although small, is not negligible and could imply a quite sizeable ladder-ladder interaction (J' in Fig. 4) that affects the nature of the low temperature magnetic state, independent magnetic measurements were performed. Thus, zero-field and longitudinal field μ^+SR measurements as a function of temperature have confirmed that [(DT-TTF)$_2$Au(mnt)$_2$] is a molecular material with a real two-leg spin ladder configuration since the results obtained corroborate the absence of a magnetic ordering, as expected for even-leg spin ladders with the quantum spin-liquid state realised [27].

There is a second example of a TTF-based organic spin ladder that has been reported by Komatsu et al. [22]. The compound, (BEDT-TTF)Zn(SCN)$_3$, belongs to the fruitful family of bis(ethylenedithio)tetrathiofulvalene (BEDT-TTF) radical ion salts (Scheme 3).

The structure of this compound consists of two parallel uniform stacks of completely ionised BEDT-TTF donors along the c-axis that are isolated from the neighbouring pairs of stacks by the diamagnetic Zn(SCN)$_3^-$ anions, as shown in Fig. 12. The (BEDT-TTF)$^{+\bullet}$ stacks form a structural two-leg ladder since they are connected by side-by-side S···S interactions (interaction 3 in Fig. 12) between BEDT-TTF molecules as well as by another weaker face-to-face interaction (interaction 2 in Fig. 12).

The low conductivity and the semiconducting behaviour of this salt denote that electrons are localised in the BEDT-TTF molecules and, therefore, the (BEDT-TTF)$^{+\cdot}$ stacks can be regarded as uniform $S = 1/2$ chains. Based on the calculated overlap integrals (Table 1) and applying Eq. (2), with an assumed value of 1.0 eV for U, the authors argue that there are four possible exchange coupling constants: one along the chain ($J_1/k = -966$ K), two interchains ($J_2/k = -5.2$ K and $J_3/k = -86.4$ K), and another, very small ($J_4/k = -0.09$ K), with an interladder character. Nevertheless, only two of these magnetic couplings can be considered as effective: those along the legs ($J_1 \equiv J_\parallel$) and those along the rungs ($J_3 \equiv J_\perp$) of the ladder. J_2, that is 1/17 of J_3, can be regarded as a very weak perturbation to the ladder interaction (Fig. 13). On the other hand, the ladders are well isolated from each other since the inter-ladder interactions J_4 are smaller that 0.1 K.

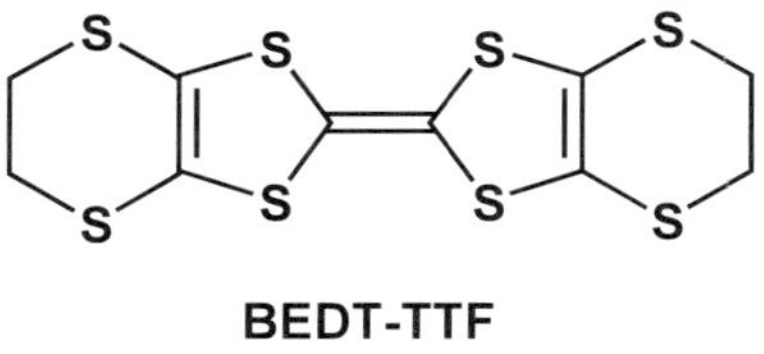

BEDT-TTF

Scheme 3.

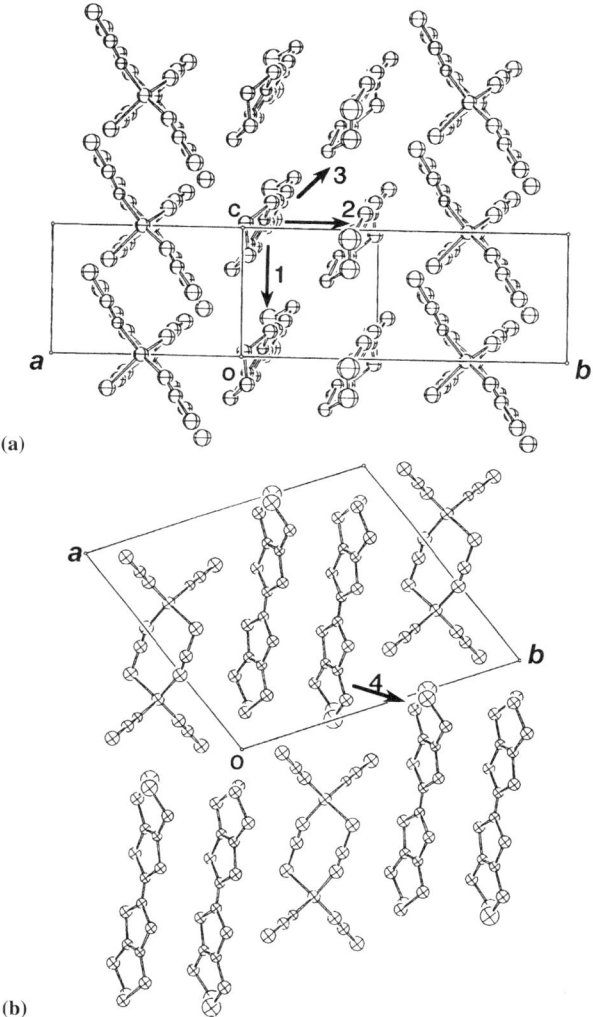

Fig. 12. X-ray structure of (BEDT-TTF)Zn(SCN)$_3$ following Komatsu et al. [22a]. (**a**) View along the molecular long axis of BEDT-TTF; (**b**) view along the c axis. Arrows indicate the interactions within (interactions 1–3) and between (interaction 4) the ladder

The fit of the experimental susceptibility data in the low temperature range (Fig. 14) to the Troyer expression [Eq. (3)] gives an energy gap of $\Delta/k = 340$ K. Nevertheless, in this case the energy gap calculated using Eq. (6) from the $J_\parallel$ and $J_\perp$ values, which were estimated from the transfer integrals, gives $\Delta/k = 4520$ K. This value shows a large discrepancy with the experimental one (more than one order of magnitude). Since Eq. (6) is derived from a perturbation of the dimer limit $J_\perp \gg J_\parallel$ [8] and the present case is in the

Table 1. The overlap integrals (S) and the exchange energy (J)

Interaction[a]	$S/10^{-3}$	Jk^{-1}/K
1	−20.4	−966
2	1.5	−5.2
3	−6.1	−86.4
4	0.2	0.09

[a] Interactions 1–4 correspond to those in Fig. 12.

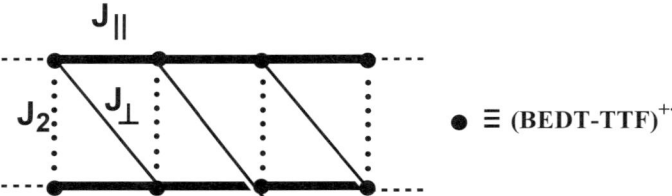

Fig. 13. Schematic representation of the two-leg spin ladder formed by the two chains of (BEDT-TTF)$^{+\bullet}$ in (BEDT-TTF)Zn(SCN)$_3$ (reprinted with permission [22a])

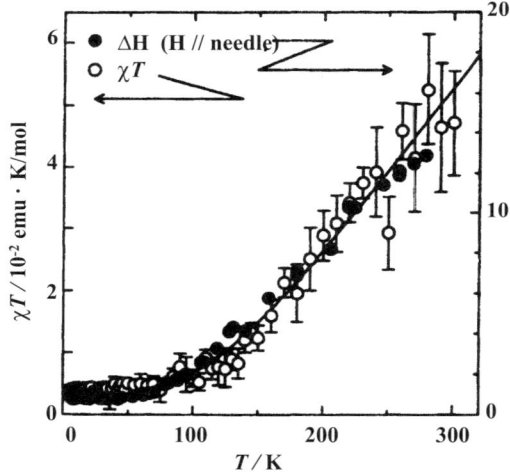

Fig. 14. Plot of χT vs. T for the (BEDT-TTF)Zn(SCN)$_3$ salt (open circles). The solid line is the best fitting curve to the spin ladder model [Eq. (3)]. Temperature dependence of ESR linewidth (closed circles) of an oriented single crystal is also shown (reprinted with permission [22a])

opposite limit of weakly interacting antiferromagnetic chains, another approximation to the spin gap, that takes into account this characteristic, was used. Such an estimation is given by a Monte Carlo calculation [28] as $\Delta = 0.41\ J_\perp$, which yields an energy gap $\Delta/k = 35$ K; a value that is one order of magnitude smaller than the experimental one. The authors stated that this result could probably be accounted for by the frustrating effect of the antiferromagnetic interaction J_2 (Fig. 13). It must be pointed out that the $J_\perp/J_\parallel$ ratio of 0.1, estimated from the transfer integrals, place this two-leg spin ladder far from an "ideal" spin ladder, more precisely at the limit in which it could also be described as pairs of weakly coupled 1D chains.

3.2
Molecular Spin Ladders by the Assembling of Dimers

The second strategy to the two-leg ladder systems – the assembly of an infinite number of dimers – has been also successful in the achievement of molecular spin ladders. By contrast with the previously described examples, in all known cases the spin carrying units are not purely organic but metal dithiolene complexes. Nevertheless, the magnetic interactions present in such complexes are mediated by overlaps between the p or π SOMO of the neighbouring spin carrying units.

Fourmigué et al. [23] have found spin ladder properties in some members of the family of charge-transfer complexes formed by the acceptor tetrafluorotetracyanoquinodimethane (TCNQF$_4$) and the flexible organometallic cyclopentadienyl/dithiolene (dithiolene = dmit^{2-}, dmid^{2-}, dsit^{2-}) Mo and W complexes as the donors (Scheme 4). These organometallic complexes are good donors forming stable radical cations that have the SOMO orbital delocalised in both the Cp$_2$M fragments and the dithiolene ligands. In every complex, the MY$_2$C$_2$ (M = Mo, W; Y = S, Se) metallacycle is folded along the Y···Y axis (see Fig. 15) and the folding angle varies strongly with the nature of the metal and the dithiolene ligand. Depending on the folding angle, the

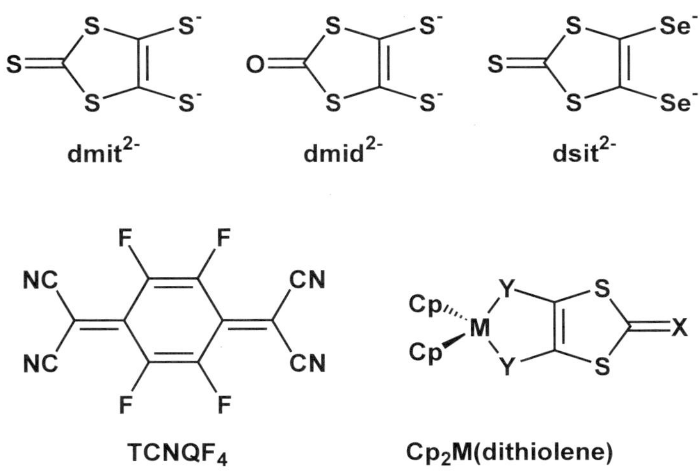

Scheme 4.

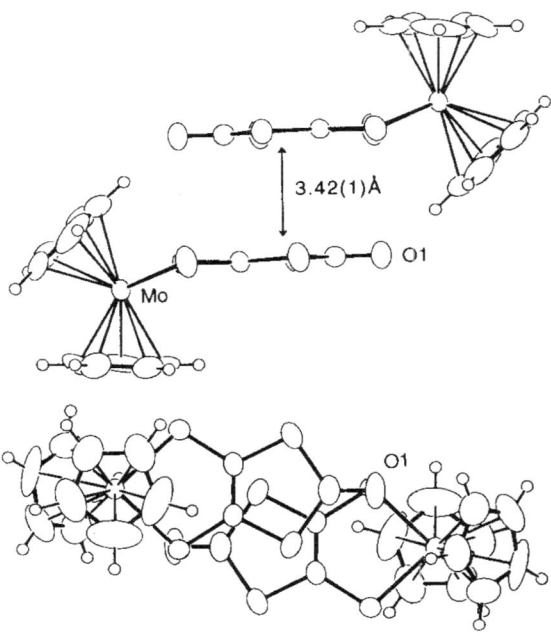

Fig. 15. Side view and top view (relative to the dmid^{2-} ligand) of the [{Cp$_2$Mo(dmid)$^{+\bullet}$}$_2$] dimer in the X-ray structure of [Cp$_2$Mo(dmid)][TCNQF$_4$] after Fourmigué et al. [23]

contribution of each fragment orbital to the SOMO varies and as a consequence so do the electronic characteristics of the building blocks.

The analysis of the X-ray structures of different compounds formed by oxidation of these donors reveal that they have a remarkable tendency to form paramagnetic [{Cp$_2$M(dithiolene)$^{+\bullet}$}$_2$] radical cation dimer motifs which interact to give different structural organisations [23, 29]. Therefore, these complexes are good building blocks to form spin ladders following the second strategy proposed if the proper interaction between them is achieved.

The strong electron acceptor TCNQF$_4$ oxidises the organometallic [Cp$_2$M(dithiolene)] donors that are organised in the resulting solid as head-to-tail centrosymmetrical dimers with parallel dithiolene planes. The reported [Cp$_2$M(dmid)][TCNQF$_4$] (M = Mo, W) complexes are isostructural being the [Cp$_2$M(dmid)]$^{+\bullet}$ radical cations dimerised by dmid/dmid π-π overlap interactions (Fig. 15). Such dimers constitute the rungs of a ladder since they are strongly interacting due to short S···S contacts, giving rise to the two-leg ladder type structures that are isolated from each other by the chains of TCNQF$_4^-$ (Fig. 16). Since TCNQF$_4^{-\bullet}$ radical anions are strongly dimerised into diamagnetic [TCNQF$_4$]$_2^{2-}$ species, only the [Cp$_2$M(dithiolene)]$^{+\bullet}$ cation radicals are responsible for the magnetic properties shown by these solids.

In accord with the described structural characteristics, the exponential decay of susceptibility data below the maximum can be fitted to the Troyer equation [Eq. (1)] giving different spin gaps for both compounds (Fig. 17);

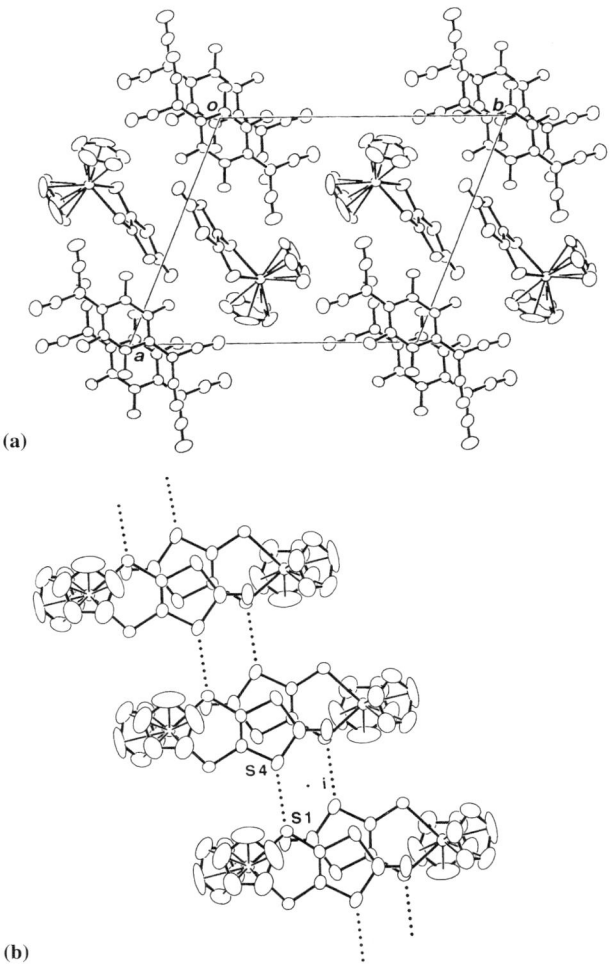

Fig. 16. (a) X-ray structure of [Cp$_2$Mo(dmid)][TCNQF$_4$] after Fourmigué et al. [23]. (b) View of the ladder structure formed by the $S = 1/2$ cations in the [Cp$_2$Mo(dmid)][TCNQF$_4$] salt (reprinted with permission [23])

$\Delta/k = 74$ K and $\Delta/k = 13$ K for the Mo and W salts, respectively. Thus, the replacement of W for Mo, while not altering the structural characteristics of the supramolecular structures, modifies the electronic characteristics of the [{Cp$_2$M(dmid)$^{+\bullet}$}$_2$] building block and as a consequence the electronic structure of the molecular spin ladder.

The authors extracted the values of $J_\parallel$ and $J_\perp$ combining the Eqs. (6) and (7). The obtained exchange coupling constants for the Mo salt were $J_\perp = 107$ K and $J_\parallel = 41$ K and for the W salt $J_\perp = 23$ K and $J_\parallel = 16$ K. The relative values of these constants are in accordance with the smaller dithiolene contribution

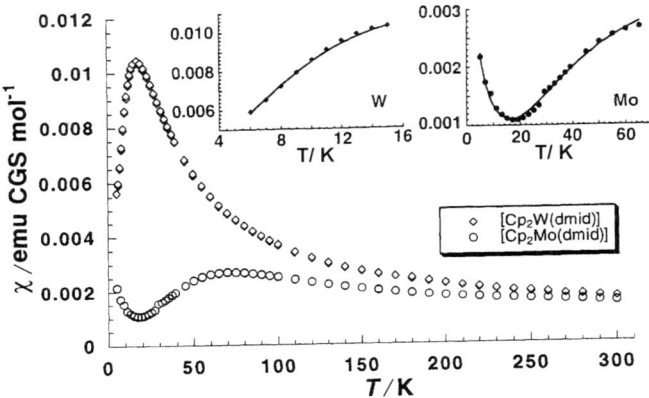

Fig. 17. Plot of magnetic susceptibility vs. T of the [Cp$_2$M(dmid)] [TCNQF$_4$] (M = W, Mo). Insets show the fits of the low temperature data to [Eq. (3)] [23]

to the SOMO for the W complex and are also in very good agreement with the calculated interaction energies within the dimers. Remarkable also are the $J_\perp/J_\parallel$ ratios of 2.6 and 1.5 for Mo and W salts, respectively, since they are close to that of an "ideal" two-leg spin ladder.

$$T(\chi_{max})/|J_\perp| = 0.625 - 0.01835 J_\parallel/J_\perp + 0.2532(J_\parallel/J_\perp)^2 \tag{7}$$

The structure of the complex [Cp$_2$W(dmit)] [TCNQF$_4$] also presents interacting dimers although with another type of dimer association, due to S···S interactions through their p orbitals instead of π-π overlaps (Fig. 18). The interaction of the dimers gives rise to a structural ladder motif (Fig. 19).

The susceptibility data (Fig. 20) were analysed as a two-leg spin ladder system, as in the other complexes, giving a spin gap value of Δ/k = 40 K [23].

The last known example of molecular spin ladder is provided by another dithiolene complex, the anionic π-radical [Ni(dmit)$_2$]$^{-\bullet}$ (Scheme 5) [24]. This planar S = 1/2 ion also forms dimeric units that interact in a ladder-like structure in the salt formed with a paramagnetic counterion, the p-N-ethylpyridinium α-nitronyl nitroxide (p-EPYNN]).

In the crystal structure of [p-EPYNN][Ni(dmit)$_2$] salt, the open-shell p-EPYNN cations form chains along c axis that isolate the chains of dimers formed also along c axis by the dithiolate radical-anions (Fig. 21). The dimers of [Ni(dmit)$_2$]$^{-\bullet}$ are formed by the ordinary plane-to-plane π-π overlap and the connection between the dimers are provided by very short (3.31 Å) S···S contacts that take place along the c axis.

In contrast to the previous [Cp$_2$M(dithiolene)] [TCNQF$_4$] compounds, the crystals of [p-EPYNN][Ni(dmit)$_2$] contain two different magnetic subsystems: the p-EPYNN cation radical chains and the one dimensional ladder-chain of [Ni(dmit)$_2$]$^{-\bullet}$ anion radicals. The chains of p-EPYNN cation radicals are arranged in very similar way to that observed in the salt [p-EPYNN][Au(dmit)$_2$] in which only the organic radicals are responsible

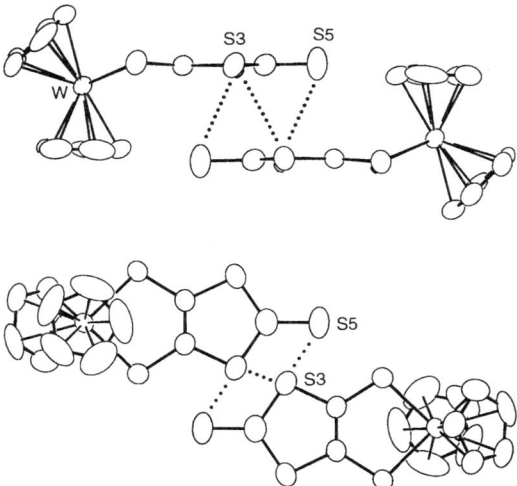

Fig. 18. Side view and top view (relative to the dmit^{2-} ligand) of the [{Cp$_2$W(dmit)$^{+\bullet}$}$_2$] dimer in the X-ray structure of [Cp$_2$W(dmit)][TCNQF$_4$] after Fourmigué et al. [23]

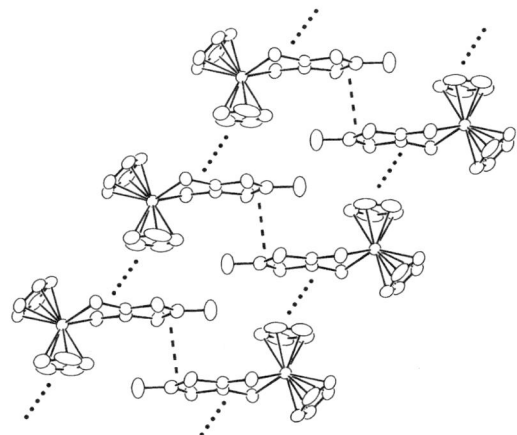

Fig. 19. View of the ladder-like structure formed by the $S = 1/2$ cations in the [Cp$_2$W(dmit)][TCNQF$_4$] salt after Fourmigué et al. [23]. Intra-dimer interactions: - - - - -, inter-dimer interactions: ······

for the magnetic properties [24b]. For this compound, the paramagnetic susceptibility can be fitted to the expression for a 1D ferromagnetic chain [1d]. The authors analysed the susceptibility data of the [p-EPYNN][Ni(dmit)$_2$] compound as originating from two independent magnetic subsystems (Fig. 22). Thus, they considered the observed two distinct temperature dependencies, one above 150 K and the other below 40 K, to originate from

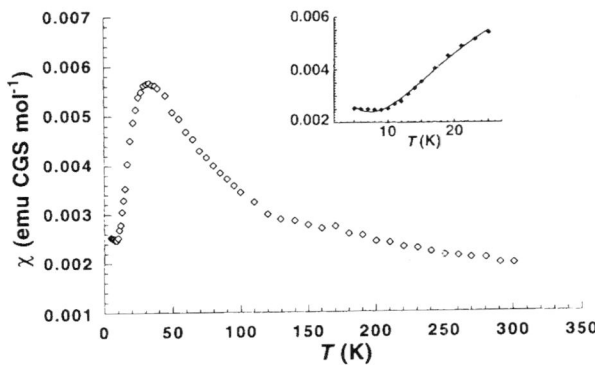

Fig. 20. Plot of magnetic susceptibility vs. T of the [Cp$_2$W(dmit)] [TCNQF$_4$]. Insert show the fit of the low temperature data to [Eq. (3)] [23]

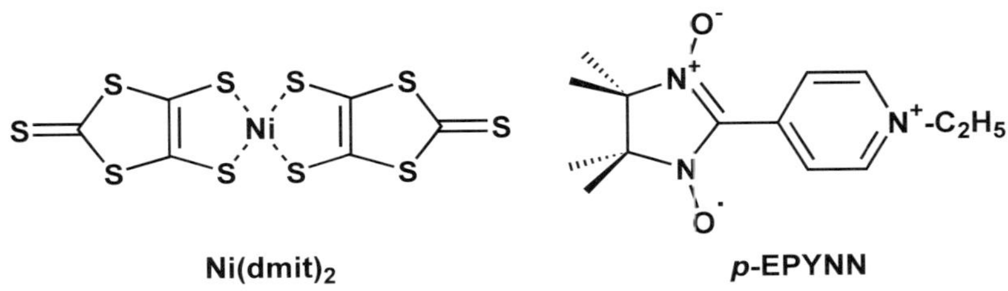

Scheme 5.

these two subsystems. The low temperature region, where an increase of $\chi_p T$ with decreasing temperature is observed, has been ascribed to the p-EPYNN cation radical chains showing a 1D ferromagnetic chain behaviour, since the behaviour is very similar to that found in the [p-EPYNN][Au(dmit)$_2$] compound (Fig. 23) which shows the same structural arrangement of radicals. The high temperature region, where there is an decrease of $\chi_p T$ with decreasing the temperature is ascribed to a spin ladder behaviour for the chain of dimers formed by the [Ni(dmit)$_2$]$^{-\bullet}$ anion radicals. The fitting of the susceptibility data shown in Fig. 22 to the sum of expressions for a spin ladder [Eq. (3)] and for a 1D chain with ferrromagnetic coupling [1d] gave a spin gap of $\Delta/k = 940$ K for the Ni(dmit)$_2$ ladder and a ferromagnetic exchange interaction of $J/k = 0.16$ K for the [p-EPYNN] chains.

Extended Huckel calculations of the overlap integrals between adjacent molecules along the legs and in the rungs gave a small value (0.13×10^{-3}) for the overlap along the [Ni(dmit)$_2$] chains and a significantly larger value (-17.0×10^{-3}) for the plane-to-plane overlapped molecules in the rungs. The evaluation, from these values, of the exchange energies $J_\perp$ and $J_\parallel$ of this system gives a $J_\perp/J_\parallel$ ratio of 10^4 that is far away from the limit of "ideal" spin ladders and is more in accordance with a non-interacting dimer description.

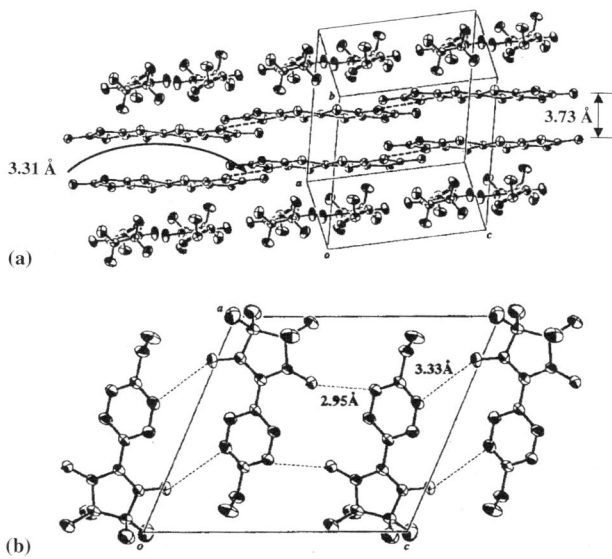

Fig. 21. Crystal structure of [p-EPYNN][Ni(dmit)$_2$] salt after Imai et al. [24b]. (a) Both chains of cations and anions are shown. Short S···S interdimer contacts are denoted by doted lines. (b) Molecular arrangement in the p-EPYNN chain

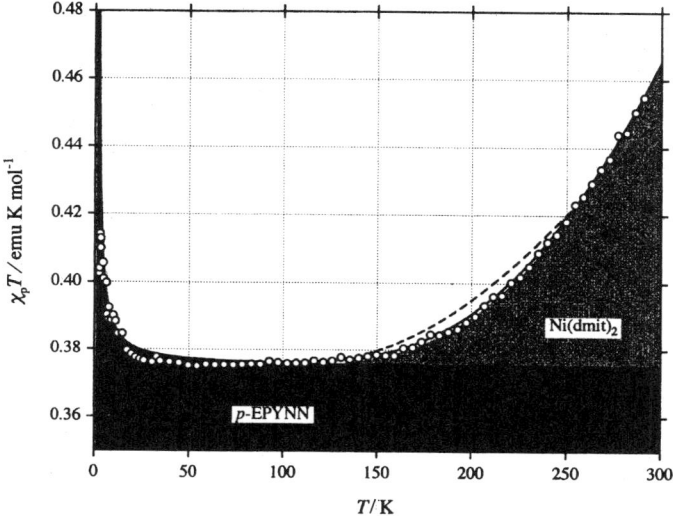

Fig. 22. Plot of susceptibility of [p-EPYNN][Ni(dmit)$_2$] vs. T [24]. The solid curve is the fit to the sum of ladder and ferromagnetic chain equations and the black and grey regions represents the contribution from [p-EPYNN] and [Ni(dmit)$_2$], respectively. The broken curve is the best fit using a simple dimer model

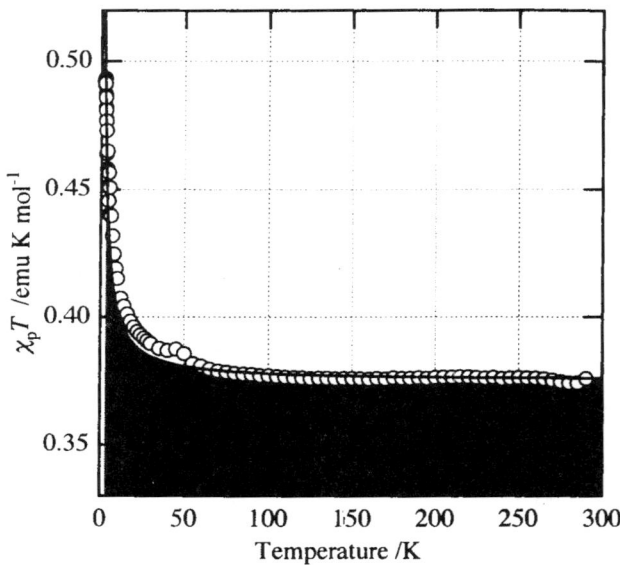

Fig. 23. Plot of susceptibility of [p-EPYNN][Au(dmit)$_2$] vs. T [24]. The solid curve is the fit to the ferromagnetic chain equation

Nevertheless, the best fitting of the susceptibility data to a simple dimer model (broken line in Fig. 22) with a $\Delta/k = 870$ K, manifests significant deviation from the experimental plots and the observed relatively large conductivity for a half-filled insulator, however, suggests that the interaction along the legs may possibly be larger than that estimated.

4
Perspectives

It is important to emphasise that although spin ladder physics has emerged only very recently as a fertile subfield of condensed matter studies, molecular spin ladders have appeared at the initation of this area of research. This fact reveals both the high level reached by the field of molecule-based magnets and the spectacular progress of supramolecular chemistry for materials development witnessed in the recent years. In fact the examples presented here, show that supramolecular chemistry is a very powerful tool to construct solid compounds with tailored physical properties such as spin ladders. Typical supramolecular and crystal engineering criteria like π-π overlap, S···S interactions and complementary nature of size and shape, used with success in the tuning of dimensionality and properties of molecular conductors and magnets have also proven to be useful in the construction of intermediate magnetic dimensionalities which are the spin ladders. Although it is not an easy task, application of these criteria will help in increasing the number of examples of molecular spin ladders that will aid in the understanding of low-dimensional spin systems exhibiting spin gaps. Another interesting aspect of the molecular spin ladders is that the interacting electrons are mainly π-electrons since they are located in molecular orbitals with a large or dominant role of s and p atomic orbitals. Important challenges in this field that deserve to be reached with molecular compounds are the construction of spin-ladders with different spin gap values able to be measured experimentally. Also important is the preparation of spin ladders with different degrees of ladder-ladder interactions and molecular spin ladders with more than two legs. Molecular compounds with such characteristics will help in the corroboration of theoretical findings about the limits of ladder-ladder exchange coupling interaction values that can destroy the ladder characteristics, as well as to verify the dependence of the bulk magnetic properties with the odd/even number of legs in the ladders.

Acknowledgements. This work was supported by grants from the DGES (BQU2000-1157), from Generalitat de Catalunya (2000SRG00114) and by the TMR Program of the EC (ERBFMRX CT980181). The author wishes to thank M. Almeida (Sacavém), J.-P. Pouget (Orsay), E. Canadell (ICMAB) and especially J. Veciana (ICMAB) and E. Ribera (ICMAB) for fruitful discussions and stimulating work.

5
References

1. For some reviews of molecular magnetic materials, see: (a) Iwamura H (1990) Adv Phys Org Chem 26: 179; (b) Day P (1993) Science 26: 431; (c) Miller JS, Epstein AJ (1994) Angew Chem Int Ed 33: 385; (d) Khan O (1993) Molecular Magnetism, VCH Publishers, New York; (e) Lathi PM (1999) Magnetic Properties of Organic Materials, Marcel Dekker, New York; (f) Day P, Underhill AE (eds) (1999) Metal-organic and organic molecular magnets. Phil Trans R Soc London A 357: 2849-3184; (g) Veciana J, Rovira C, Amabilino DB (1999) Supramolecular Engineering of Synthetic Metallic Materials: Conductors and Magnets, Kluwer, Dordrecht, The Netherlands
2. Dagotto E, Riera J, Scalapino DJ (1992) Phys Rev B 45: 5744
3. (a) Dagotto E, Rice TM (1996) Science 271: 618; (b) Scalapino DJ (1995) Nature 377: 12; (c) Hiroi Z, Takano M (1995) Nature 337: 41
4. (a) Rice TM, Goplan S, Sigrist M (1993) Europhys Lett 23: 445; (b) Barnes T, Dagotto E, Riera J, Swanson ES (1993) Phys Rev B 47: 3196
5. Goplan S, Rice TM, Sigrist M (1994) Phys Rev B 49: 8901
6. Noack RM, White SR, Scalapino DJ (1994) Phys Rev Lett 73: 882
7. Sigrist M, Rice TM, Zhang FC (1994) Phys Rev B 49: 1058
8. Troyer M, Tsunetsugu H, Würtz D (1994) Phys Rev B 50: 13515
9. (a) Chaboussant G, Crowell PA, Lévy LP, Piovesana O, Madouri A, Mailly D (1997) Phys Rev B 55: 3046; (b) Chaboussant G, Julien M-H, Fagot-Revurat Y, Lévy LP, Berthier C, Horvatié M, Piovesana O (1997) Phys Rev Lett 79: 925; (c) Elstner N, Singh RRP (1998) Phys Rev B 58: 11484
10. Jeckelmann E, Scalapino DJ, White SR (1998) Phys Rev B 58: 9492
11. Troyer M, Zhitomirsky ME, Ueda K (1997) Phys Rev B 55: 6117
12. Azuma M, Hiroi Z, Tanako M, Ishida K, Kitaoka I (1994) Phys Rev Lett 73: 3463
13. (a) Johnston DC, Johnson JW, Goshom DP, Jacobson AP (1987) Phys Rev B 35: 219; (b) Barnes T, Riera J (1994) Phys Rev B 49: 6817

14. Hiroi Z, Takano M (1995) Nature 377: 41
15. Kojima K, Keren A, Luke GM, Nachumi B, Wu WD, Uemura YJ, Azuma M, Tanako M (1995) Phys Rev Lett 74: 2812
16. Kadono R, Okajima H, Yamasita A, Yokoo T, Akimutsu J, Kobayashi N, Hiroi Z, Takano M, Nagamine K (1996) Phys Rev B 54: 9628
17. Matsumoto S, Kitaoka Y, Ishida K, Asayama K, Hiroi Z, Kobayashi N, Takano M (1996) Phys Rev B 53: 11942
18. Garret AW, Nagler SE, Tennant DA, Sales BC, Barnes T (1997) Phys Rev Lett 79: 74
19. (a) Uehara U, Nagata T, Akimitsu J, Takahashi H, Mori N, Kinoshita K (1996) J Phys Soc Jpn 65: 2764; (b) Szymczak R, Szymczak H, Baran M, Mosiniewicz-Szablewska E, Leonyuk L, Babonas G-J, Maltsev V, Shvanskaya L (1999) Physica C 311: 187; (c) Katano S, Nagata T, Akimitsu J, Nishi M, Kakurai K (1999) Phys Rev Lett 18: 636
20. (a) Jérome D (1991) Science 252: 1509; (b) Vescoli V, Degiorgi L, Henderson W, Grüner G, Starkey KP, Montgomery LK (1998) Science 281: 1181; (c) Bourbonnais C, Jérome D (1998) Science 281: 1155; (d) Bernier P, Lefrant S, Bidan G (1999) Advances in Synthetic Metals. Twenty Years of Progress in Science and Technology Elsevier, Lausanne; (e) Conwell E (1988) Highly Conducting Quasi-One-Dimensional Organic Crystals, Semiconductors and Semimetals, Vol 27 Academic Press, Inc London; (f) Saito G, Kagoshima S (1990) The Physics and Chemistry of Organic Superconductors, Springer, Berlin; (g) Williams JM, Ferraro JR, Thorn RJ, Carlson KD, Geiser U, Wang HH, Kini AM, Whangbo MH (1992) Organic Superconductors, Including Fullerenes: Synthesis, Structure, Properties, and Theory Prentice-Hall, Englewood Cliffs, NJ
21. (a) Rovira C, Veciana J, Ribera E, Tarrés J, Candell E, Rousseau R, Mas M, Molins E, Almeida M, Henriques RT, Morgado J, Schoeffel JP, Pouget JP (1997) Angew Chem Int Ed Engl 36: 2324; (b) Rovira C, Veciana J, Ribera E, Tarres J, Candell E, Rousseau R, Mas M, Molins E, Almeida M, Henriques RT, Morgado J, Schoeffel JP, Pouget JP (1999) Chem Eur J 5: 2025
22. (a) Komatsu T, Kojima N, Saito G (1997) Solid State Commun 103: 519; (b) Komatsu T, Kojima N, Saito G (1999) Synthetic Metals 103: 1923
23. Fourmigué M, Domerq B, Jourdain IV, Molinié P, Guyon F, Amaudrut J (1998) Chem Eur J 4: 1714
24. (a) Imai H, Inabe T, Otsuka T, Okumo T, Awaga K (1996) Phys Rev B 54: R6338; (b) Imai H, Otsuka T, Naito T, Agawa K, Inabe T (1999) J Am Chem Soc 121: 8098
25. Novoa JJ, Rovira MC, Rovira C, Veciana J, Tarrés J (1995) Adv Mat 7: 233
26. Rovira C, Veciana J, Santaló N, Tarrés J, Cirujeda J, Molins E, Llorca J, Espinosa E (1994) J Org Chem 59: 3307
27. Arcon D, Lappas A, Margadonna S, Prassides K, Ribera E, Veciana J, Rovira C, Henriques RT, Almeida M (1999) Phys Rev B 60: 4191
28. Graven M, Birgeneau RJ, Wiese U-J (1996) 77: 1865
29. Fourmigué M, Domercq B (1999) In: Veciana J, Rovira C, Amabilino DB (eds) Supramolecular engineering of Synthetic Metallic Materials. Kluwer Academic Publishers, Netherlands, p 329

Author Index Volumes 1–100

Abolmaali B, Taylor HV, Weser U (1998) Evolutionary Aspects of Copper Binding Centres in Copper Proteins. *91*: 91–190
Adam W, Mitchell CM, Saha-Möller CR, Weichhold O (2000) Structure, Reactivity and Selectivity of Metal–Peroxo Complexes Versus Dioxiranes. *97*: 237–286
Aegerter MA (1996) Sol–Gel Chromogenic Materials and Devices. *85*: 149–194
Ahrland S (1966) Factors Contributing to (b)-behavior in Acceptors. *1*: 207–220
Ahrland S (1968) Thermodynamics of Complex Formation between Hard and Soft Acceptors and Donors. *5*: 118–149
Ahrland S (1973) Thermodynamics of the Stepwise Formation of Metal-Ion Complexes in Aqueous Solution. *15*: 167–188
Aisen P, see Doi K (1980) *70*: 1–26
Alcock NW, see Leciejewicz J (1995) *82*: 43–84
Allan CB, see Maroney MJ (1998) *92*: 1–66
Allen GC, Warren KD (1971) The Electronic Spectra of the Hexafluoro Complexes of the First Transition series. *9*: 49–138
Allen GC, Warren KD (1974) The Electronic Spectra of the Hexafluoro Complexes of the Second and Third Transition Series. *19*: 105–165
Alonso JA, Balbas LC (1993) Hardness of Metallic Clusters. *80*: 229–258
Alonso JA, Balbás LC (1987) Simple Density Functional Theory of the Electronegativity and Other Related Properties of Atoms and Ions. *66*: 41–78
Amendola V (2001) Molecular Movements and Translocations Controlled by Transition Metals and Signalled by Light Emission. *99*: 79–115
Andersson LA, Dawson JH (1991) EXAFS Spectroscopy of Heme-Containing Oxygenases and Peroxidases. *74*: 1–40
Antanaitis BC, see Doi K (1988) *70*: 1–26
Arčon D, see Prassides K (2001) *100*: 129–162
Ardon M, Bino A (1987) A New Aspect of Hydrolysis of Metal Ions: The Hydrogen Oxide Bridging Ligand (H_3O_2). *65*: 1–28
Arendsen F, see Hagen WR (1998) *90*: 161–192
Armstrong FA (1990) Probing Metalloproteins by Voltammetry. *72*: 137–221
Athanassopoulou MA, see also Haase W (1999) *94*: 139–197
Augustynski J (1988) Aspects of Photo-Electrochemical and Surface Behavior of Titanium(IV) Oxide. *69*: 1–61
Auld DS (1997) Zinc Catalysis in Metalloproteases. *89*: 29–50
Averill BA (1983) FE—S and Mo—Fe—S Clusters as Models for the Active Site of Nitrogenase. *53*: 57–101

Babel D (1967) Structural Chemistry of Octahedral Fluorocomplexes of the Transition Elements. *3*: 1–87
Bacci M (1984) The Role of Vibronic Coupling in the Interpretation of Spectroscopic and Structural Properties of Biomolecules. *55*: 67–99

Baekelandt BG, Mortier WJ, Schonheydt RA (1993) The EEM Approach to Chemical Hardness in Molecules and Solids: Fundamentals and Applications. 80: 187–228
Baker EC, Halstead GW, Raymond KN (1976) The Structure and Bonding of 4f and 5f Series Organometallic Compounds. 25: 21–66
Balbás LC, see Alonso JA (1987) 66: 41–78
Balbás LC, see Alono JA (1993) 80: 229–258
Baldwin AH, see Butler A (1997) 89: 109–132
Ballardini R (2001) Molecular-Level Artificial Machines Based on Photoinduced Electron-Transfer Processes. 99: 163–188
Balsenc LR (1980) Sulfur Interaction with Surfaces and Interfaces Studied by Auger Electron Spectrometry. 39: 83–114
Balzani V, see Ballardini R (2001) 99: 163–188
Banci L, Bencini A, Benelli C, Gatteschi D, Zanchini C (1982) Spectral–Structural Correlations in High-Spin Cobalt(II) Complexes. 52: 37–86
Banci L, Bertini I, Luchinat C (1990) The ^{1}H NMR Parameters of Magnetically Coupled Dimers – The Fe_2S_2 Proteins as an Example. 72: 113–136
Banse F, see Girerd JJ (2000) 97: 145–178
Baran EJ, see Müller A (1976) 26: 81–139
Bartolotti LJ (1987) Absolute Electronegativities as Determined from Kohn-Sham Theory. 66: 27–40
Bates MA, Luckhurst GR (1999) Computer Simulation of Liquid Crystal Phases Formed by Gay-Berne Mesogens. 94: 65–137
Bau RG, see Teller R (1981) 44: 1–82
Baughan EC (1973) Structural Radii, Electron-cloud Radii, Ionic Radii and Solvation. 15: 53–71
Bayer E, Schretzmann P (1967) Reversible Oxygenierung von Metallkomplexen. 2: 181–250
Bearden AJ, Dunham WR (1970) Iron Electronic Configuration in Proteins: Studies by Mössbauer Spectroscopy. 8: 1–52
Bencini A, see Banci L (1982) 52: 37–86
Benedict C, see Manes L (1985) 59/60: 75–125
Benelli C, see Banci L (1982) 52: 37–86
Benfield RE, see Thiel RC (1993) 81: 1–40
Bergmann D, Hinze J (1987) Electronegativity and Charge Distribution. 66: 145–190
Bernadou J, see Meunier B (2000) 97: 1–36
Berners-Price SJ, Sadler PJ (1988) Phosphines and Metal Phosphine Complexes: Relationship of Chemistry to Anticancer and Other Biological Activity. 70: 27–102
Bernt I, see Uller E (2000) 96: 149–176
Bertini I, see Banci L (1990) 72: 113–136
Bertini I, Ciurli S, Luchinat C (1995) The Electronic Structure of FeS Centérs in Proteins and Models. A Contribution to the Understanding of Their Electron Transfer Properties. 83: 1–54
Bertini I, Luchinat C, Scozzafava A (1982) Carbonic Anhydrase: An Insight into the Zinc Binding Site and into the Active Cavity Through Metal Substitution. 48: 45–91
Bertrand P (1991) Application of Electron Transfer Theories to Biological Systems. 75: 1–48
Bill E, see Trautwein AX (1991) 78: 1–96
Bino A, see Ardon M (1987) 65: 1–28
Blackman AG, see Tolman WB (2000) 97: 179–210
Blanchard M, see Linares C (1977) 33: 1
Blasse G, see Powell RC (1980) 42: 43–96
Blasse G (1991) Optical Electron Transfer Between Metal Ions and its Consequences. 76: 153–188
Blasse G (1976) The Influence of Charge-Transfer and Rydberg States on the Luminescence Properties of Lanthanides and Actinides. 26: 43–79
Blasse G (1980) The Luminescence of Closed-Shell Transition Metal-Complexes. New Developments. 42: 1–41

Blauer G (1974) Optical Activity of Conjugated Proteins. *18*: 69–129
Bleijenberg KC (1980) Luminescence Properties of Uranate Centres in Solids. *42*: 97–128
Boca R, Breza M, Pelikán P (1989) Vibronic Interactions in the Stereochemistry of Metal Complexes. *71*: 57–97
Boeyens JCA (1985) Molecular Mechanics and the Structure Hypothesis. *63*: 65–101
Bögge H, see Müller A (2000) *96*: 203–236
Böhm MC, see Sen KD (1987) *66*: 99–123
Bohra R, see Jain VK (1982) *52*: 147–196
Bollinger DM, see Orchin M (1975) *23*: 167–193
Bominaar EL, see Trautwein AX (1991) *78*: 1–96
Bonnelle C (1976) Band and Localized States in Metallic Thorium, Uranium and Plutonium and in some compounds, Studied by X-ray Spectroscopy. *31*: 23–48
Bose SN, see Nag K (1985) *63*: 153–197
Bowler BE, see Therien MJ (1991) *75*: 109–130
Bradshaw AM, Cederbaum LS, Domcke W (1975) Ultraviolet Photoelectron Spectroscopy of Gases Adsorbed on Metal Surfaces. *24*: 133–170
Braterman PS (1972) Spectra and Bonding in Metal Carbonyls. Part A: Bonding. *10*: 57–86
Braterman PS (1976) Spectra and Bonding in Metal Carbonyls. Part B: Spectra and Their Interpretation. *26*: 1–42
Bray RC, Swann JC (1972) Molybdenum-Containing Enzymes. *11*: 107–144
Brec R, see Evain M (1992) *79*: 277–306
Brese NE, O'Keeffe M (1992) Crystal Chemistry of Inorganic Nitrides. *79*: 307–378
Breza M, see Boca R (1989) *71*: 57–97
Briggs LR, see Kustin K (1983) *53*: 137–158
Brooks MSS (1985) The Theory of 5f Bonding in Actinide Solids. *59/60*: 263–293
Brown DG, see Wood JM (1972) *11*: 47–105
Bruce DW, see also Donnio B (1999) *95*: 193–247
Buchanan BB (1966) The Chemistry and Function of Ferredoxin. *1*: 109–148
Bucher E, see Campagna M (1976) *30*: 99–140
Buchler JW, Dreher C, Künzel FM (1995) Synthesis and Coordination Chemistry of Noble Metal Porphyrins. *84*: 1–70
Buchler JW, Kokisch W, Smith PD (1978) Cis, Trans, and Metal Effects in Transition Metal Porphyrins. *34*: 79–134
Bulman RA (1978) Chemistry of Plutonium and the Transuranics in the Biosphere. *34*: 39–77
Bulman RA (1987) The Chemistry of Chelating Agents in Medical Sciences. *67*: 91–141
Burdett JK (1987) Some Structural Problems Examined Using the Method of Moments. *65*: 29–90
Burdett JK (1976) The Shapes of Main-Group Molecules: A Simple Semi-Quantitative Molecular Orbital Approach. *31*: 67–105
Burger RM (2000) Nature of Activated Bleomycin. *97*: 287–304
Burgmayer SJN (1998) Electron Transfer in Transition Metal–Pteridine Systems. *92*: 67–119
Butler A, Baldwin AH (1997) Vanadium Bromoperoxidase and Functional Mimics. *89*: 109–132

Campagna M, Wertheim GK, Bucher E (1976) Spectroscopy of Homogeneous Mixed Valence Rare Earth Compounds. *30*: 99–140
Capozzi F, Ciurli S, Luchinat C (1998) Coordination Sphere Versus Protein Environment as Determinants of Electronic and Functional Properties of Iron-Sulfur Proteins *90*: 127–160
Carr AJ, see Melendez RE (2000) *96*: 31–62
Carter RO, see Müller A (1976) *26*: 81–139
Cauletti C, see Furlani C (1978) *35*: 119–169
Cederbaum LS, see Bradshaw AM (1975) *24*: 133–170
Cederbaum LS, see Schmelcher PS (1996) *86*: 27–62
Ceulemans A, Vanquickenborne LG (1989) The Epikernel Principle. *71*: 125–159

Chandrasekhar V, Thomas KR, Justin KR (1993) Recent Aspects of the Structure and Reactivity of Cyclophosphazenes. 81: 41–114
Chandrashekar TK, see Ravikanth M (1995) 82: 105–188
Chang J, see Therien MJ (1991) 75: 109–130
Chapman SK, Daff S, Munro AW (1997) Heme: The Most Versatile Redox Centre in Biology? 88: 39–70
Chasteen ND (1983) The Biochemistry of Vanadium. 53: 103–136
Chattaraj PK, Parr RG (1993) Density of Functional Theory of Chemical Hardness. 80: 11–26
Cheh AM, Neilands JP (1976) The γ-Aminoevulinate Dehydratases: Molecular and Environmental Properties. 29: 123–169
Chimiak A, Neilands JB (1984) Lysine Analogues of Siderophores. 58: 89–96
Christensen JJ, see Izatt RM (1973) 16: 161–189
Ciampolini M (1969) Spectra of 3d Five-Coordinate Complexes. 6: 52–93
Ciurli S, see Bertini I (1995) 83: 1–54
Ciurli S, see Capozzi F (1998) 90: 127–160
Clack DW, Warren KD (1980) Metal-Ligand Bonding in 3d Sandwich Complexes. 39: 1–141
Clark SJ, see also Crain J (1999) 94: 1–39
Clarke MJ, Fackler PH (1982) The Chemistry of Technetium: Toward Improved Diagnostic Agents. 50: 57–58
Clarke MJ, Gaul JB (1993) Chemistry Relevant to the Biological Effects of Nitric Oxide and Metallonitrosyls. 81: 147–181
Clarke RJH, Stewart B (1979) The Resonance Raman Effect. Review of the Theory and of Applications in Inorganic Chemistry. 36: 1–80
Codling K, Frasinski LJ (1996) Molecules in Intense Laser Fields: an Experimental Viewpoint. 86: 1–26
Cohen IA (1980) Metal-Metal Interactions in Metalloporphyrins, Metalloproteins and Metalloenzymes. 40: 1–37
Connett PH, Wetterhahn KE (1983) Metabolism of the Carcinogen Chromate by Cellular Constituents. 54: 93–124
Cook DB (1978) The Approximate Calculation of Molecular Electronic Structures as a Theory of Valence. 35: 37–86
Cooper SL (2001) Optical Spectroscopic Studies of Metal–Insulator Transitions in Perovskite-Related Oxides. 98: 161–220
Cooper SR, Rawle SC (1990) Crown Thioether Chemistry. 72: 1–72
Corbett JD (1997) Diverse Naked Clusters of the Heavy Main-Group Elements. Electronic Regularities and Analogies. 87: 157–194
Corbin PS, see Zimmerman SC (2000) 96: 63–94
Cotton FA, Walton RA (1985) Metal–Metal Multiple Bonds in Dinuclear Clusters. 62: 1–49
Cox PA (1975) Fractional Parentage Methods for Ionisation of Open Shells of d and f Electrons. 24: 59–81
Cox MC, see Sun H (1997) 88: 71–102
Crabtree RH, see Siegbahn PEM (2000) 97: 125–144
Crain J, Clark SJ (1999) Calculation of Structure and Dynamical Properties of Liquid Crystal. 94: 1–39
Cras JA, see Willemse J (1976) 28: 83–126
Credi A, see Ballardini R (2001) 99: 163–188
Cremer D, see Frenking G (1990) 73: 17–96
Crichton RR (1973) Ferritin. 17: 67–134

Daff S, see Chapman SK (1997) 88: 39–70
Dance J-M, see Tressaud A (1982) 52: 87–146
Darriet J, see Drillon M (1992) 79: 55–100
Daul C, Schläpfer CW, von Zelewsky A (1979) The Electronic Structure of Cobalt(II) Complexes with Schiff Bases and Related Ligands. 36: 129–171

Davidson G, see Maroney MJ (1998) 92: 1–66
Davidson P (1999) Selected Topics in X-Ray Scattering by Liquid–Crystalline Polymers. 95: 1–39
Dawson JH, see Andersson LA (1991) 74: 1–40
Deeth RJ (1995) Computational Modelling of Transition Metal Centres. 82: 1–42
Degen J, see Schmidtke H-H (1989) 71: 99–124
Dehnicke K, Shihada A-F (1976) Structural and Bonding Aspects in Phosphorus Chemistry – Inorganic Derivates of Oxohalogeno Phosphoric Acids. 28: 51–82
Demleitner B, see Uller E (2000) 96: 149–176
Denning RG (1992) Electronic Structure and Bonding in Actinyl Ions. 79: 215–276
Deumal M, see Novoa JJ (2001) 100: 33–60
Dhubhghaill OMN, Sadler PJ (1991) The Structure and Reactivity of Arsenic Compounds. Biological Activity and Drug Design. 78: 129–190
Diehn B, see Doughty MJ (1980) 41: 45–70
Diemann E, see Müller A (1973) 14: 23–47
Dirken MW, see Thiel RC (1993) 81: 1–40
Dobiás B (1984) Surfactant Adsorption on Minerals Related to Flotation. 56: 91–147
Doi K, Antanaitis BC, Aisen P (1998) The Binuclear Iron Centers of Uteroferrin and the Purple Acid Phosphatases. 70: 1–26
Domcke W, see Bradshaw AM (1975) 24: 133–170
Donnio B, Bruce DW (1999) Metallomesogens 95: 193–247
Dophin D, see Morgan B (1987) 64: 115–204
Doughty MJ, Diehn B (1980) Flavins as Photoreceptor Pigments for Behavioral Responses. 41: 45–70
Drago RS (1973) Quantitative Evaluation and Prediction of Donor–Acceptor Interactions. 15: 73–139
Dreher C, see Buchler JW (1995) 84: 1–70
Drillon M, Darriet J (1992) Progress in Polymetallic Exchange-Coupled Systems, some Examples in Inorganic Chemistry. 79: 55–100
Duffy JA (1977) Optical Electronegativity and Nephelauxetic Effect in Oxide Systems. 32: 147–166
Dunham WR, see Bearden AJ (1970) 8: 1–52
Dunn MF (1975) Mechanisms of Zinc Ion Catalysis in Small Molecules and Enzymes. 23: 61–122

Ealtough DJ, see Izatt RM (1973) 16: 161–189
Egami T (2001) Local Atomic Structure of CMR Manganites and Related Oxides. 98: 115–160
Eller PG, see Ryan RR (1981) 46: 47–100
Emmerling A, see Fricke J (1991) 77: 37–88
Emsley E (1984) The Composition, Structure and Hydrogen Bonding of the β-Diketones. 57: 147–191
Englman R (1981) Vibrations in Interaction with Impurities. 43: 113–158
Epstein IR, Kustin K (1984) Design of Inorganic Chemical Oscillators. 56: 1–33
Ermer O (1976) Calculations of Molecular Properties Using Force Fields. Applications in Organic Chemistry. 27: 161–211
Ernst RD (1984) Structure and Bonding in Metal-Pentadienyl and Related Compounds. 57: 1–53
Erskine RW, Field BO (1976) Reversible Oxygenation. 28: 1–50
Evain M, Brec R (1992) A new Approach to Structural Description of Complex Polyhedra Containing Polychalcogenide Anions. 79: 277–306

Fabbrizzi L, see Amendola V (2001) 99: 79–115
Fackler PH, see Clarke MJ (1982) 50: 57–58
Fajans K (1967) Degrees of Polarity and Mutual Polarization of Ions in the Molecules of Alkali Fluorides, SrO and BaO. 3: 88–105

Fan M-F, see Lin Z (1997) 87: 35–80
Fee JA (1975) Copper Proteins – Systems Containing the "Blue" Copper Center. 23: 1–60
Feeney RE, Komatsu SK (1966) The Transferrins. 1: 149–206
Fehlner TP (1997) Metalloboranes. 87: 111–136
Felsche J (1973) The Crystal Chemistry of the Rare-Earth Silicates. 13: 99–197
Ferreira R (1976) Paradoxical Violations of Koopmans' Theorem, with Special Reference to the 3d Transition Elements and the Lanthanides. 31: 1–21
Fichtinger-Schepman AMJ, see Reedijk J (1987) 67: 53–89
Fidelis IK, Mioduski T (1981) Double–Double Effect in the Inner Transition Elements. 47: 27–51
Field BO, see Erskine RW (1976) 28: 1–50
Figlar J, see Maroney MJ (1988) 92: 1–66
Fischer J, see Mathey F (1984) 55: 153–201
Fischer S, see Tytho KH (1999) 93: 125–317
Follmann H, see Lammers M (1983) 54: 27–91
Fonticella-Camps JC (1998) Biological Nickel. 91: 1–30
Fournier JM, Manes L (1985) Actinide Solids. 5f Dependence of Physical Properties. 59/60: 1–56
Fournier JM (1985) Magnetic Properties of Actinide Solids. 59/60: 127–196
Fraga S, Valdemoro C (1968) Quantum Chemical Studies on the Submolecular Structure of the Nucleic Acids. 4: 1–62
Frasinski LJ, see Codling K (1996) 85: 1–26
Fraústo da Ailva JJR, Williams RJP (1976) The Uptake of Elements by Biological Systems. 29: 67–121
Frenking G, see Jørgensen CK (1990) 73: 1–16
Frenking G, Cremer D (1990) The Chemistry of the Noble Gas Elements Helium, Neon, and Argon – Experimental Facts and Theoretical Predictions. 73: 17–96
Frey M (1998) Nickel-Iron Hydrogenases: Structural and Functional Properties. 90: 97–126
Fricke B (1975) Superheavy Elements. 21: 89–144
Fricke J, Emmerling A (1991) Aerogels-Preparation, Properties, Applications. 77: 37–88
Friebel C, see Reinen D (1979) 37: 1–60
Friedrich H (1996) Field Induced Chao and Chaotic Scattering. 86: 97–124
Friesen C, see Keppler BK (1991) 78: 97–128
Fuhrhop J-H (1974) The Oxidation States and Reversible Redox Reactions of Metalloporphyrins. 18: 1–67
Fujii H, see Watanabe Y (2000) 97: 61–90
Fujita M (2000) Molecular Paneling Through Metal-Directed Self-Assembly. 96: 177–202
Furlani C, Cauletti C (1978) He(1) Photo Electron Spectra of d-metal Compounds. 35: 119–169

Gani D, Wilkie J (1997) Metal Ions in the Mechanism of Enzyme Catalysed Phosphate Monoester Hydrolyses. 89: 133–176
Gallagher TF (1996) Microwave Multiphoton Exitation and Ionization. 86: 125–148
Galland P, see Russo VEA (1980) 41: 71–110
Galván M, see Gázquez JL (1987) 66: 79–98
Gandolfi MT, see Ballardini R (2001) 99: 163–188
Gatteschi D, see Banci L (1982) 52: 37–86
Gaul JB, see Clarke MJ (1993) 81: 147–181
Gavezzotti A, see Simonetta M (1976) 27: 1–43
Gázquez JL, Vela A, Galván M (1987) Fukui Function, Electronegativity and Hardness in the Kohn-Sham Theory. 66: 79–98
Gazquéz JL (1993) Hardness and Softness in Density Functional Theory. 80: 27–44
Gerloch M, Harding JH, Woolley RG (1981) The Context and Application of Ligand Field Theory. 46: 1–46
Ghijsen J, see Naegele JR (1985) 59/60: 197–262

Gilert TR, see Kustin K (1983) 53: 137-158
Gillard RD, Mitchell PR (1970) The Absolute Configuration of Transition Metal Complexes. 7: 46-86
Gimzewski J, see Joachim C (2001) 99: 1-18
Girerd JJ, Banse F, Simaan AJ (2000) Characterization and Properties of Non-Heme Iron-Peroxo Complexes. 97: 145-178
Gleitzer C, Goodenough JB (1985) Mixed-Valence Iron Oxides. 61: 1-76
Gliemann G, Yersin H (1985) Spectroscopic Properties of the Quasi One-Dimensional Tetracyanoplatinate(II) Compounds. 62: 87-153
Golovina AP, Zorov NB, Runov VK (1981) Chemical Luminescence Analysis of Inorganic substances. 47: 53-119
Gómez-Kaifer M, see Liu J (2001) 99: 141-162
Goodby JW (1999) Twist Grain Boundary (TGB) Phases. 95: 83-147
Goodenough JB, see Gleitzer C (1985) 61: 1-76
Goodenough JB (2001) General Considerations. 98: 1-16
Goodenough JB (2001) Transport Properties. 98: 17-114
Grätzel M, see Kiwi J (1982) 49: 37-125
Gray HB, see Therien MJ (1991) 75: 109-130
Green JC (1981) Gas Phase Photoelectron Spectra of d- and f-Block Organometallic Compounds. 43: 37-112
Grenier JC, Pouchard M, Hagenmuller P (1981) Vacancy Ordering in Oxygen-Deficient Perovskite-Related Ferrites. 47: 1-25
Grice ME, see Politzer P (1993) 80: 101-114
Griffith JS (1972) On the General Theory of Magnetic Susceptibilities of Polynuclear Transitionmetal Compounds. 10: 87-126
Grisham CM, see Mildvan AS (1974) 20: 1-21
Gubelmann MH, Williams AF (1984) The Structure and Reactivity of Dioxygen Complexes of the Transition Metals. 55: 1-65
Güdel HU, see Ludi A (1973) 14: 1-21
Guilard R, Lecomte C, Kadish KM (1987) Synthesis, Electrochemistry, and Structural Properties of Porphyrins with Metal-Carbon Single Bonds and Metal-Metal Bonds. 64: 5-268
Guillaumont R, see Hubert S (1978) 34: 1-18
Guillon D (1999) Columnar Order in Thermotropic Mesophases. 95: 41-82
Gütlich P (1981) Spin Crossover in Iron(II)-Complexes. 44: 83-195
Gutmann V, see Mayer U (1972) 12: 113-140
Gutmann V, Mayer U (1973) Redox Properties: Changes Effected by Coordination. 15: 141-166
Gutmann V, Mayer U (1972) Thermochemistry of the Chemical Bond. 10: 127-151
Gutmann V, Mayer H (1976) Application of the Functional Approach to Bond Variations Under Pressure. 31: 49-66

Haase W, Athanasspoulou MA (1999) Crystal Structure of LC Mesogens. 94: 139-197
Häder D-P, see Nultsch W (1980) 41: 111-139
Hagen WR, Arendsen AF (1998) The Bio-Inorganic Chemistry of Tungsten. 90: 161-192
Hagenmuller P, see Grenier JC (1981) 47: 1-25
Hale JD, see Williams RJP (1966) 1: 249-281
Hale JD, see Williams RJP (1973) 15: 1 and 2
Halet J-F, Saillard J-Y (1997) Electron Count Versus Structural Arrangement in Clusters Based on a Cubic Transition Metal Core with Bridging Main Group Elements. 87: 81-110
Hall DI, Ling JH, Nyholm RS (1973) Metal Complexes of Chelating Olefin-Group V Ligands. 15: 3-51
Halstead GW, see Baker EC (1976) 25: 21-66
Hamilton AD, see Meléndez RE (2000) 96: 31-62
Hamstra BJ, see Slebodnick C (1997) 89: 51-108
Hanack M, see Schultz H (1991) 74: 41-146

Harding JH, see Gerloch M (1981) *46*: 1–46
Harnung SE, Schäffer CE (1972) Phase-fixed 3-G Symbols and Coupling Coefficients for the Point Groups. *12*: 257–255
Harnung SE, Schäffer CE (1972) Real Irreducible Tensorial Sets and their Applications to the Ligand-Field Theory. *12*: 257–295
Harris WR (1998) Binding and Transport of Nonferrous Metal by Serum Transferrin. *92*: 121–162
Hathaway BJ (1984) A New Look at the Stereochemistry and Electronic Properties of Complexes of the Copper(II) Ion. *57*: 55–118
Hathaway BJ (1973) The Evidence for "Out-of-the Plane" Bonding in Axial Complexes of the Copper(II) Ion. *14*: 49–67
Hawes JC, see Mingos DMP (1985) *63*: 1–63
Hellner EE (1979) The Frameworks (Bauverbände) of the Cubic Structure Types. *37*: 61–140
Hemmerich P, Michel H, Schung C, Massey V (1982) Scope and Limitation of Single Electron Transfer in Biology. *48*: 93–124
Henry M, Jolivet JP, Livage J (1991) Aqueous Chemistry of Metal Cations: Hydrolysis, Condensation and Complexation. *77*: 153–206
Herrmann WA, see Kühn FE (2000) *97*: 213–236
Hider RC (1984) Siderophores Mediated Absorption of Iron. *57*: 25–88
Hill HAO, Röder A, Williams RJP (1970) The Chemical Nature and Reactivity of Cytochrome P-450. *8*: 123–151
Hilpert K (1990) Chemistry of Inorganic Vapors. *73*: 97–198
Hinze J, see Bergmann D (1987) *66*: 145–190
Hoffman BM, Natan MJ, Nocek JM, Wallin SA (1991) Long-Range Electron Transfer Within Metal-Substituted Protein Complexes. *75*: 85–108
Hoffmann BM, see Ibers JA (1982) *50*: 1–55
Hoffmann DK, Ruedenberg K, Verkade JG (1977) Molecular Orbital Bonding Concepts in Polyatomic Molecules – A Novel Pictorial Approach. *33*: 57–96
Hogenkamp HPC, Sando GN (1974) The Enzymatic Reduction of Ribonucleotides. *20*: 23–58
Housecroft CE (1997) Clusters with Interstitial Atoms from the p-Block: How Do Wade's Rules Handle Them? *87*: 137–156
Huber R, see Ramao MJ (1998) *90*: 69–96
Hubert S, Hussonois M, Guillaumont R (1978) Measurement of Complexing Constants by Radiochemical Methods. *34*: 1–18
Hudson RF (1966) Displacement Reactions and Concept of Soft and Hard Acids and Bases. *1*: 221–223
Hulliger F (1968) Crystal Chemistry of Chalcogenides and Pnictides of the Transition Elements. *4*: 83–229
Hussonois M, see Hubert S (1978) *34*: 1–18
Hyde BG, see Makovicky E (1981) *46*: 101–170
Hyde BG, see O'Keeffe M (1985) *61*: 77–144

Ibers JA, Pace LJ, Martinsen J, Hoffmann BM (1982) Stacked Metal Complexes: Structures and Properties. *50*: 1–55
Imrie CT (1999) Liquid Crystal Dimers. *95*: 149–192
Ingraham LL, see Maggiora GM (1967) *2*: 126–159
Inoue K (2001) Nitroxide Radical-Metal-Based Molecular Magnets. *100*: 61–92
Iqbal Z (1972) Intra- and Inter-Molecular Bonding and Structure of Inorganic Pseudohalides with Triatomic Groupings. *10*: 25–55
Izatt RM, Eatough DJ, Christensen JJ (1973) Thermodynamics of Cation–Macrocyclic Compound Interaction. *16*: 161–189

Jain VK, Bohra R, Mehrotra RC (1982) Structure and Bonding in Organic Derivatives of Antimony(V). *52*: 147–196

Jerome–Lerutte S (1972) Vibrational Spectra and Structural Properties of Complex Tetra cyanides of Platinum, Palladium and Nickel. *10*: 153–166
Joachim C (2001) Single Molecular Rotor at the Nanoscale. *99*: 1–18
Johnston RL (1997) Mathematical Cluster Chemistry. *87*: 1–34
Johnston RL, see Mingos DMP (1987) *68*: 29–87
Jolivet JP, see Henry M (1991) *77*: 153–206
Jørgensen CK, see Müller A (1973) *14*: 23–47
Jørgensen CK, see Reisfeld R (1982) *49*: 1–36
Jørgensen CK, see Reisfeld R (1988) *69*: 63–96
Jørgensen CK, see Reisfeld R (1991) *77*: 207–256
Jørgensen CK, Frenking G (1990) Historical, Spectroscopic and Chemical Comparison of Noble Gases. *73*: 1–16
Jørgensen CK, Kauffmann GB (1990) Crookes and Marignac – A Centennial of an Intuitive and Pragmatic Appraisal of "Chemical Elements" and the Present Astrophysical Status of Nucleosynthesis and "Dark Matter". *73*: 227–254
Jørgensen CK, Reisfeld R (1982) Uranyl Photophysics. *50*: 121–171
Jørgensen CK (1976) Deep-Lying Valence Orbitals and Problems of Degeneracy and Intensitities in Photo-Electron Spectra. *30*: 141–192
Jørgensen CK (1966) Electric Polarizability, Innocent Ligands and Spectroscopic Oxidation States. *1*: 234–248
Jørgensen CK (1990) Heavy Elements Synthesized in Supernovae and Detected in Peculiar A-type Stars. *73*: 199–226
Jørgensen CK (1996) Luminescence of Cerium(III) Inter-Shell Transitions and Scintillator Action. *85*: 195–214
Jørgensen CK (1976) Narrow Band Thermoluminescence (Candoluminescence) of Rare Earths in Auer Mantles. *25*: 1–20
Jørgensen CK (1975) Partly Filled Shells Constituting Anti–bonding Orbitals with Higher Ionization Energy than Their Bonding Counterparts. *22*: 49–81
Jørgensen CK (1975) Photo-Electron Spectra of Non-Metallic Solids and Consequences for Quantum Chemistry. *24*: 1–58
Jørgensen CK (1978) Predictable Quarkonium Chemistry. *34*: 19–38
Jørgensen CK (1966) Recent Progress in Ligand Field Theory. *1*: 3–31
Jørgensen CK (1967) Relationship Between Softness, Covalent Bonding, Ionicity and Electric Polarizability. *3*: 106–115
Jørgensen CK (1981) The Conditions for Total Symmetry Stabilizing Molecules, Atoms, Nuclei and Hadrons. *43*: 1–36
Jørgensen CK (1973) The Inner Mechanism of Rare Earths Elucidated by Photo-Electron Spectra. *13*: 199–253
Jørgensen CK (1969) Valence-Shell Expansion Studied by Ultra-violet Spectroscopy. *6*: 94–115
Justin KR, see Chandrasekhar V (1993) *81*: 41–114

Kadish KM, see Guilard R (1987) *64*: 205–268
Kahn O (1987) Magnetism of the Heteropolymetallic Systems. *68*: 89–167
Kaifer AE, see Liu J (2001) *99*: 141–162
Kalyanasundaram K, see Kiwi J (1982) *49*: 37–125
Kato T (2000) Hydrogen-Bonded Liquid Crystals – Molecular Self-Assembly for Dynamically Functional Materials. *96*: 95–146
Katz E, see Shipway AN (2001) *99*: 237–281
Kauffmann GB, see Jørgensen CK (1990) *73*: 227–254
Keijzers CP, see Willemse J (1976) *28*: 83–126
Kelly JM, see Moucheron C (1998) *92*: 163–216
Kelly TR (2001) Rotary Motion in Single-Molecule Machines. *99*: 19–53
Kemp TJ, see Leciejewicz J (1995) *82*: 43–84

Keppler BK, Friesen C, Moritz HG, Vongerichten H, Vogel E (1991) Tumor-Inhibiting Bis (β-Diketonato) Metal Complexes. Budotitane, cis-Diethoxybis (1-phenylbutane-1, 3-dionato) titanium(IV). 78: 97–128

Kimura E, Koike T, Shionoya M (1997) Advances in Zinc Enzyme Models by Small, Mononuclear Zinc(II) Complexes. 89: 1–28

Kimura T (1968) Biochemical Aspects of Iron Sulfur Linkage in Non-Heme Iron Protein, with Special Reference to "Adrenodoxin". 5: 1–40

Kinoshita M (2001) An Organic Radical Crystal Showing Spontaneous Ferromagnetic Order. 100: 1–32

Kirsch-De Mesmaekar A, see Moucheron C (1998) 92: 163–216

Kitagawa T, Ozaki Y (1987) Infrared and Raman Spectra of Metalloporphyrins. 64: 71–114

Kiwi J, Kalyanasundaram K, Grätzel M (1982) Visible Light Induced Cleavage of Water into Hydrogen and Oxygen in Colloidal and Microheterogeneous Systems. 49: 37–125

Kjekshus A, Rakke T (1974) Considerations of the Valence Concept. 19: 45–83

Kjekshus A, Rakke T(1974) Geometrical Considerations on the Marcasite Type Structure. 19: 85–104

Klabunde T, Krebs B (1997) The Dimetal Center in Purple Acid Phosphatases. 89: 177–198

Kögerler P, see Müller A (2000) 96: 203–236

Koike T, see Kimura E (1997) 89: 1–28

Kokisch W, see Buchler JW (1978) 34: 79–134

Komatsu SK, see Feeney RE (1966) 1: 149–206

Komorowski L (1993) Hardness Indices for Free and Bonded Atoms. 80: 45–70

König E (1991) Nature and Dynamics of the Spin-State Interconversions in Metal Complexes. 76: 51–152

König E (1971) The Nephelauxetic Effect. Calculation and Accuracy of the Interelectronic Repulsion Parameters 1. Cubic High-Spin d_2, d_3, d_7 and d_8 Systems. 9: 175–212

Köpf H, see Köpf-Maier P (1988) 70: 103–185

Köpf-Maier P, Köpf H (1988) Transition and Main-Group Metal Cyclopentadienyl Complexes: Preclinical Studies on a Series of Antitumor Agents of Different Structural Type: 70: 103–185

Koppikar DK, Sivapullaiah PV, Ramakrishnan L, Soundararajan S (1978) Complexes of the Lanthanides with Neutral Oxygen Donor Ligands. 34: 135–213

Kóren B, see Valach F (1984) 55: 101–151

Krause R (1987) Synthesis of Ruthenium (II) Complexes of Aromatic Chelating Heterocycles: Towards the Design of Luminescent Compounds. 67: 1–52

Krebs B, see Klabunde T (1997) 89: 177–198

Krische MJ, Lehn JM (2000) The Utilization of Persistent H-Bonding Motifs in the Self-Assembly of Supramolecular Architectures. 96: 3–30

Krumholz P (1971) Iron(II) Diimine and Related Complexes. 9: 139–174

Kühn FE, Herrmann WA (2000) Rhenium-Oxo and Rhenium-Peroxo Complexes in Catalytic Oxidations. 97: 213–236

Kubas GJ, see Ryan RR (1981) 46: 47–100

Kuki A (1991) Electronic Tunneling Paths in Proteins. 75: 49–84

Kulander KC, Schafer KJ (1996) Time-Dependent Calculations of Electron and Photon Emission from an Atom in an Intense Laser Field. 86: 149–172

Künzel FM, see Buchler JW (1995) 84: 1–70

Kurad D, see Tytko KH (1999) 93: 1–64

Kustin K, see Epstein IR (1984) 56: 1–33

Kustin K, McLeod GC, Gilbert TR, Briggs LR (1983) Vanadium and Other Metal Ions in the Physiological Ecology of Marine Organisms. 53: 137–158

Labarre JF (1978) Conformational Analysis in Inorganic Chemistry: Semi-Empirical Quantum Calculation vs. Experiment. 35: 1–35

Lammers M, Follmann H (1983) The Ribonucleotide Reductases: A Unique Group of Metallo-enzymes Essential for Cell Proliferation. 54: 27–91

Le Brun NE, Thomson AJ, Moore GR (1997) Metal Centres of Bacterioferritins or Non-Heam-Iron-Containing Cytochromes b_{557}. 88: 103–138

Leciejewicz J, Alcock NW, Kemp TJ (1995) Carboxylato Complexes of the Uranyl Ion: Effects Ligand Size and Coordinat. Geometry Upon Molecular and Crystal Structure. 82: 43–84

Lecomte C, see Guilard R (1987) 64: 205–268

Lee YJ, see Scheidt WR (1987) 64: 1–70

Lehmann H, see Schultz H (1991) 74: 41–146

Lehn J-M (1973) Design of Organic Complexing Agents. Strategies Towards Properties. 16: 1–59

Lehn JM, see Krische MJ (2000) 96: 3–30

Li H, see Sun H (1997) 88: 71–102

Lioccia S, Paolesse R (1995) Metal Complexes of Corroles and Other Corrinoids. 84: 71–134

Lin Z, Fan M-F (1997) Metal–Metal Interactions in Transition Metal Clusters with π-Donor Ligands. 87: 35–80

Linarés C, Louat A, Blanchard M (1977) Rare-Earth Oxygen Bonding in the LnMO$_4$ Xenotime Structure. 33: 179–207

Lindskog S (1970) Cobalt(II) in Metalloenzymes. A Reporter of Structure–Function Relations. 8: 153–196

Ling JH, see Hall DI (1973) 15: 3–51

Linton BR, see Meléndez RE (2000) 96: 31–62

Liu A, Neilands JB (1984) Mutational Analysis of Rhodotorulic Acid Synthesis in Rhodotorula philimanae. 58: 97–106

Liu J (2001) Switchable Molecular Devices: From Rotaxanes to Nanoparticles. 99: 141–162

Livage J, see Henry M (1991) 77: 153–206

Livorness J, Smith T (1982) The Role of Manganese in Photosynthesis. 48: 1–44

Llinás M (1973) Metal–Polypeptide Interactions: The Conformational State of Iron Proteins. 17: 135–220

Louat A, see Linarés C (1977) 33: 179–207

Luchinat C, see Banci L (1990) 72: 113–136

Luchinat C, see Bertini I (1982) 48: 45–91

Luchinat C, see Bertini I (1995) 83: 1–54

Luchinat C, see Capozzi F (1998) 90: 127–160

Lucken EAC (1969) Valence-Shell Expansion Studied by Radio-Frequency Spectroscopy. 6: 1–29

Luckhurst GR, see also Bates MA (1999) 94: 65–137

Ludi A, Güdel HU (1973) Structural Chemistry of Polynuclear Transition Metal Cyanides. 14: 1–21

Lutz HD (1988) Bonding and Structure of Water Molecules in Solid Hydrates. Correlation of Spectroscopic and Structural Data. 69: 125

Lutz HD (1995) Hydroxide Ions in Condensed Materials – Correlation of Spectroscopy and Structural Data. 82: 85–104

Maaskant WJA (1995) On Helices Resulting from a Cooperative Jahn-Teller Effect in Hexagonal Perovskites. 83: 55–88

Maggiora GM, Ingraham LL (1967) Chlorophyll Triplet States. 2: 126–159

Magyar B (1973) Salzebullioskopie III. 14: 111–140

Makovicky E, Hyde BG (1981) Non-Commensurate (Misfit) Layer Structures. 46: 101–170

Manes L, see Fournier JM (1985) 59/60: 1–56

Manes L, Benedict U (1985) Structural and Thermodynamic Properties of Actinide Solids and Their Relation to Bonding. 59/60: 75–125

Mangano C, see Amendola V (2001) 99: 79–115

Mann S (1983) Mineralization in Biological Systems. 54: 125–174

March NH (1993) The Ground-State Energy of Atomic and Molecular Ions and Its Variation with the Number of Electrons. 80: 71–86

March NH (1996) Semiclassical Theory of Atoms and Ions in Intense External Fields. 86: 63–96

Maroney MJ, Davidson G, Allan CB, Figlar J (1998) The Structure and Function of Nickel Sites in Metalloproteins. 92: 1–66

Martinsen J, see Ibers JA (1982) 50: 1–55

Mason SF (1980) The Ligand Polarization Model for the Spectra of Metal Complexes: The Dynamic Coupling Transition Probabilities. 39: 43–81

Massey V, see Hemmerich P (1982) 48: 93–124

Mathey F, Fischer J, Nelson JH (1984) Complexing Modes of the Phosphole Moiety. 55: 153–201

Mauk AG (1991) Electron Transfer in Genetically Engineered Proteins. The Cytochrome c Paradigm. 75: 131–158

Mayer U, see Gutman V (1972) 10: 127–151

Mayer U, see Gutman V (1973) 15: 141–166

Mayer H, see Gutman V (1976) 31: 49–66

Mayer U, Gutman V (1972) Phenomenological Approach to Cation–Solvent Interactions. 12: 113–140

Mazumdar S, Mitra S (1993) Biomimetic Chemistry of Hemes Inside Aqueous Micelles. 81: 115–145

McGrady JE, see Mingos DMP (1992) 79: 1–54

McLendon G (1991) Control of Biological Electron Transport via Molecular Recognition and Binding: The "Velcro" Model. 75: 159–174

McLeod GC, see Kustin K (1983) 53: 137–158

Mehmke J, see Tytko KH (1999) 93: 1–64

Mehmke J, see Tytko KH (1999) 93: 125–317

Mehrotra RC, see Jain VK (1982) 52: 147–196

Mehrotra RC (1991) Present Status and Future Potential of the Sol-Gel Process. 77: 1–36

Meier PC, see Simon W (1973) 16: 113–160

Meléndez RE, Carr AJ, Linton BR, Hamilton AD (2000) Controlling Hydrogen Bonding: From Molecular Recognition to Organogelation. 96: 31–62

Melnik M, see Valach F (1984) 55: 101–151

Messerschmidt A (1998) Metal Sites in Small Blue Copper Proteins, Blue Copper Oxidase and Vanadium-Containing Enzymes. 90: 37–68

Meunier B, Bernadou J (2000) Active Iron-Oxo and Iron-Peroxo Species in Cytochromes P-450 and Peroxidases; Oxo-Hydroxo Tautomerism with Water-Soluble Metalloporphyrins. 97: 1–36

Michel H, see Hemmerich P (1982) 48: 93–124

Mildvan AS, Grishan CM (1974) The Role of Divalent Cations in the Mechanism of Enzyme Catalyzed Phosphoryl and Nucleotidyl. 20: 1–21

Mingos DMP, Hawes JC (1985) Complementary Spherical Electron Density Model. 63: 1–63

Mingos DMP, Johnston RL (1987) Theoretical Models of Cluster Bonding. 68: 29–87

Mingos DMP, McGrady JE, Rohl AL (1992) Moments of Inertia in Cluster and Coordination Compounds. 79: 1–54

Mingos DMP, Zhenyang L (1990) Hybridization Schemes for Coordination and Organometallic Compounds. 72: 73–112

Mingos DMP, Zhenyang L (1989) Non-Bonding Orbitals in Coordination Hydrocarbon and Cluster Compounds. 71: 1–56

Mioduski T, see Fidelis IK (1981) 47: 27–51

Mitchell CM, see Adam W (2000) 97: 237–286

Mitchell PR, see Gillard RD (1970) 7: 46–86

Mitra S, see Mazumdar S (1993) 81: 115–145

Moody DC, see Ryan RR (1981) 46: 47–100

Moore GR, see Le Brun NE (1997) 88: 103–138

Moreau-Colin ML (1972) Electronic Spectra and Structural Properties of Complex Tetracyanides of Platinum, Palladium and Nickel. *10*: 167–190
Morf WE, see Simon W (1973) *16*: 113–160
Morgan B, Dophin D (1987) Synthesis and Structure of Biometric Porphyrins. *64*: 115–204
Moritz HG, see Keppler BK (1991) *78*: 97–128
Morris DFC (1968/1969) An Appendix to Structure and Bonding. *4*; *6*: 157–159
Morris DFC (1968) Ionic Radii and Enthalpies of Hydration of Ions. *4*: 63–82
Mortensen OS (1987) A Noncommuting-Generator Approach to Molecular Symmetry. *68*: 1–28
Mortier JW (1987) Electronegativity Equalization and its Application. *66*: 125–143
Mortier WJ, see Baekelandt BG (1993) *80*: 187–228
Moucheron C, Kirsch-De Mesmaeker A, Kelly JM (1998) Photophysics and Photochemistry of Metal Polypyridyl and Related Complexes with Nucleic Acids. *92*: 163–216
Moura I, see Xavier AV(1981) *43*: 187–213
Moura JJG, see Xavier AV (1981) *43*: 187–213
Mullay JJ (1987) Estimation of Atomic and Group Electronegativities. *66*: 1–25
Müller A, Baran EJ, Carter RO (1976) Vibrational Spectra of Oxo-, Thio-, and Selenometallates of Transition Elements in the Solid State. *26*: 81–139
Müller A, Diemann F, Jørgensen CK (1973) Electronic Spectra of Tetrahedral Oxo, Thio and Seleno Complexes. Formed by Elements of the Beginning of the Transition Groups. *14*: 23–47
Müller A, Kögerler P, Bögge H (2000) Pythagorean Harmony in the World of Metal Oxygen Clusters of the $\{MO_{11}\}$ Type: Giant Wheels and Spheres both Based on Pentagonal Type Unit. *96*: 203–236
Müller U (1973) Strukurchemie der Azide. *14*: 141–172
Müller W, Spirlet J-C (1985) The Preparation of High Purity Actinide Metals and Compounds. *59/60*: 57–73
Munro AW, see Chapman SK (1997) *88*: 39–70
Murray JS, see Politzer P (1993) *80*: 101–114
Murrell JM (1977) The Potential Energy Surfaces of Polyatomic Molecules. *32*: 93–146

Naegele JR, Ghijsen J (1985) Localization and Hybridization of 5f States in the Metallic and Ionic Bond as Investigated by Photoelectron Spectroscopy. *59/60*: 197–262
Nag K, Bose SN (1985) Chemistry of Tetra- and Pentavalent Chromium. *63*: 153–197
Nalewajski RF (1993) The Hardness Based Molecular Charge Sensitivities and Their Use in the Theory of Chemical Reactivity. *80*: 115–186
Natan MJ, see Hoffman BM (1991) *75*: 85–108
Neilands JB, see Liu A (1984) *58*: 97–106
Neilands JB, see Chimiak A (1984) *58*: 89–96
Neilands JB (1972) Evolution of Biological Iron Binding Centres. *11*: 145–170
Neilands JB (1984) Methodology of Siderophores. *58*: 1–24
Neilands JB (1966) Naturally Occurring Non-Porphyrin Iron Compounds. *1*: 59–108
Neilands JP, see Cheh AM (1976) *29*: 123–169
Nelson JH, see Mathey F (1984) *55*: 153–201
Nickerson DP, see Wong L-L (1997) *88*: 175–208
Nieboer E (1975) The Lanthanide Ions as Structural Probes in Biological and Model Systems. *22*: 1–47
Nieter Burgmeier SJ (1998) Electron Transfer in Transition Metal-Pteridine Systems. *92*: 67–120
Nocek JM, see Hoffman BM (1991) *75*: 85–108
Nomoto K, see Sugiura Y (1984) *58*: 107–135
Novack A (1974) Hydrogen Bonding in Solids. Correlation of Spectroscopic and Crystallographic Data. *18*: 177–216
Novoa JJ (2001) The Mechanism of the Through-Space Magnetic Interactions in Purely Organic Molecular Magnets. *100*: 33–60

Nultsch W, Häder D-P (1980) Light Perception and Sensory Transduction in Photosynthetic Prokaryotes. *41*: 111–139
Nyholm RS, see Hall DI (1973) *15*: 3–51

O'Keeffe M, see Brese NE (1992) *79*: 307–378
O'Keeffe M, Hyde BG (1985) An Alternative Approach to Non-Molecular Crystal Structures with Emphasis on the Arrangements of Cations. *61*: 77–144
O'Keeffe M (1989) The Prediction and Interpretation of Bond Lengths in Crystals. *71*: 161–190
Odom JD (1983) Selenium Biochemistry. Chemical and Physical Studies. *54*: 1–26
Oehme I, see Wolfbeis OS (1996) *85*: 51–98
Oelkrug D (1971) Absorption Spectra and Ligand Field Parameters of Tetragonal 3d-Transition Metal Fluorides. *9*: 1–26
Oosterhuis WT (1974) The Electronic State of Iron in Some Natural Iron Compounds: Determination by Mössbauer and ESR Spectroscopy. *20*: 59–99
Orchin M, Bollinger DM (1975) Hydrogen-Deuterium Exchange in Aromatic Compounds. *23*: 167–193
Ostrovskii PI (1999) Packing and Molecular Conformation, and Their Relationship with LC Phase Behaviour. *94*: 199–240
Ozaki Y, see Kitaagawa T (1987) *64*: 71–114

Pace LJ, see Ibers JA (1982) *50*: 1–55
Padhye SB, see West DC (1991) *76*: 1–50
Palacio F, see Rawson JM (2001) *100*: 93–128
Pallavicini P, see Amendola V (2001) *99*: 79–115
Paolesse R, see Licoccia S (1995) *84*: 71–134
Parr RG, see Chattaraj PK (1993) *80*: 11–26
Patil SK, see Ramakrishna VV (1984) *56*: 35–90
Peacock RD (1975) The Intensities of Lanthanide $f \leftrightarrow f$ Transitions. *22*: 83–122
Pearson RG (1993) Chemical Hardness – An Historial Introduction. *80*: 1–10
Pease AR (2001) Computing at the Molecular Level. *99*: 189–236
Pecoraro VL, see Slebodnick C (1997) *89*: 51–108
Pelikán P, see Boca R (1989) *71*: 57–97
Penfield KW, see Solomon EI (1983) *53*: 1–56
Penneman RA, Ryan RR, Rosenzweigh A (1973) Structural Systematics in Actinide Fluoride Complexes. *13*: 1–52
Penner-Hahn JE (1998) Structural Characterization of the Mn Site in the Photosynthetic Oxygen-Evolving Complex. *90*: 1–36
Pereira IAC, Teixeira M, Xavier AV (1998) Hemeproteins in Anaerobes. *91*: 65–90
Perlman ML, see Watson RE (1975) *24*: 83–132
Politzer P, Murray JS, Grice ME (1993) Charge Capacities and Shell Structures of Atoms. *80*: 101–114
Pouchard M, see Grenier JC (1981) *47*: 1–25
Powell AK (1997) Polyiron Oxides, Oxyhydroxides and Hydroxides as Models for Biomineralisation Processes. *88*: 1–38
Powell RC, Blasse G (1980) Energy Transfer in Concentrated Systems. *42*: 43–96
Prassides K (2001) Magnetism in Fullerene Derivatives. *100*: 129–162

Que Jr. L (1980) Non-Heme Iron Dioxygenases. Structure and Mechanism. *40*: 39–72

Raehm L (2001) Molecular Machines and Motors Based on Transition Metal-Containing Catenanes and Rotaxanes *99*: 55–78
Rakke T, see Kjekshus A (1974) *19*: 45–83
Rakke T, see Kjekshus A (1974) *19*: 85–104
Ramakrishna VV, Patil SK (1984) Synergic Extraction of Actinides. *56*: 35–90

Ramakrishnan L, see Koppikar DK (1978) *34*: 135–213
Rao VUS, see Wallace WE (1977) *33*: 1–55
Raphael AL, see Therien MJ (1991) *75*: 109–130
Ravikanth M, Chandrashekar TK (1995) Nonplanar Porphyrins and Their Biological Relevance: Ground and Excited State Dynamics. *82*: 105–188
Rawle SC, see Cooper SR (1990) *72*: 1–72
Rawson JM (2001) Magnetic Properties of Thiazyl Radicals. *100*: 93–128
Raymond KN, see Baker EC (1976) *25*: 21–66
Raymond KN, Smith WL (1981) Actinide-Specific Sequestering Agents and Decontamination Applications. *43*: 159–186
Reedijk J, Fichtinger-Schepman AMJ, Oosterom AT van, Putte P van de (1987) Platinum Amine Coordination Compounds as Anti-Tumour Drugs. Molecular Aspects of the Mechanism of Action. *67*: 53–89
Rein M, see Schultz H (1991) *74*: 41–146
Reinen D, Friebel C (1979) Local and Cooperative Jahn-Teller Interactions in Model Structures. Spectroscopic and Structural Evidence. *37*: 1–60
Reinen D (1970) Kationenverteilung zweiwertiger 3dn-Ionen in oxidischen Spinell-, Granat und anderen Strukturen. *7*: 114–154
Reinen D (1969) Ligand-Field Spectroscopy and Chemical Bonding in Cr3+-Containing Oxidic Solids. *6*: 30–51
Reisfeld R, see Jørgensen CK (1982) *50*: 121–171
Reisfeld R, Jørgensen CK (1988) Excited States of Chromium(III) in Translucent Glass-Ceramics as Prospective Laser Materials. *69*: 63–96
Reisfeld R (1996) Laser Based on Sol–Gel Technology. *85*: 215–234
Reisfeld R, Jørgensen CK (1982) Luminescent Solar Concentrators for Energy Conversion. *49*: 1–36
Reisfeld R (1996) New Materials for Non-linear Optics. *85*: 99–148
Reisfeld R, Jørgensen CK (1991) Optical Properties of Colorants or Luminescent Species in Sol-Gel Classes. *77*: 207–256
Reisfeld R (1976) Excited States and Energy Transfer from Donor Cations to Rare Earths in the Condensed Phase. *30*: 65–97
Reisfeld R (1975) Radiative and Non-Radiative Transitions of Rare Earth Ions in Glasses. *22*: 123–175
Reisfeld R (1973) Spectra and Energy Transfer of Rare Earths in Inorganic Glasses. *13*: 53–98
Reisfeld R, see Wolfbeis OS (1996) *85*: 51–98
Reslova S, see Thomson AJ (1972) *11*: 1–46
Röder A, see Hill HAO (1970) *8*: 123–151
Rohl AL, see Mingos DMP (1992) *79*: 1–54
Romao MJ, Huber R (1998) Structure and Function of the Xanthine-Oxidase Family of Molybdenum Enzymes. *90*: 69–96
Rosenzweig A, see Pennemen RA (1973) *13*: 1–52
Rovira C (2001) Molecular Compounds Showing a Spin Ladder Behavior. *100*: 163–188
Rüdiger W (1980) Phytochrome, a Light Receptor of Plant Photomorphogenesis. *40*: 101–140
Ruedenberg K, see Hoffmann DK (1977) *33*: 57–96
Runov VK, see Golovina AP (1981) *47*: 53–119
Russo VEA, Galland P (1980) Sensory Physiology of Phycomyces Blakesleeanus. *41*: 71–110
Ryan RR, see Penneman RA (1973) *13*: 1–52
Ryan RR, Kubas GJ, Moody DC, Eller PG (1981) Structure and Bonding of Transition Metal-Sulfur Dioxide Complexes. *46*: 47–100

Saalfrank RW, see Uller E (2000) *96*: 149–176
Sadler PJ, see Berners-Price SJ (1988) *70*: 27–102
Sadler PJ, see Dhubhghaill OMN (1991) *78*: 129–190
Sadler PJ, see Sun H (1997) *88*: 71–102

Sadler PJ (1976) The Biological Chemistry of Gold: A Metallo-Drug and Heavy-Atom Label with Variable Valency. 29: 171–214
Saha-Möller CR, see Adam W (2000) 97: 237–286
Saillard J-Y, see Halet J-F (1997) 87: 81–110
Sakka S, Yoko T (1991) Sol–Gel-Derived Coating Films and Applications. 77: 89–118
Sakka S (1996) Sol–Gel Coating Films for Optical and Electronic Application. 85: 1–50
Saltman P, see Spiro G (1969) 6: 116–156
Sando GN, see Hogenkamp HPC (1974) 20: 23–58
Sankar SG, see Wallace WE (1977) 33: 1–55
Sano M (2001) Molecular Hysteresisby Linkage Isomerizations Induced by Electrochemical. 99: 117–139
Sauvage JP, see Raehm L (2001) 99: 55–78
Schäffer CE, see Harnung SE (1972) 12: 201–255
Schäffer CE, see Harnung SE (1972) 12: 257–295
Schäffer CE (1968) A Perturbation Representation of Weak Covalent Bonding. 5: 68–95
Schäffer CE (1973) Two Symmetry Parameterizations of the Angular-Overlap Model of the Linda-Field. Relation to the Crystal-Field Model. 14: 69–110
Scheidt WR, Lee YJ (1987) Recent Advances in the Stereochemistry of Metallotetrapyrroles. 64: 1–70
Schläpfer CW, see Daul C (1979) 36: 129–171
Schmelcher PS, Cederbaum LS (1996) Two Interacting Charged Particles in Strong Static Fields: A Variety of Two-Body Phenomena. 86: 27–62
Schmid G (1985) Developments in Transition Metal Cluster Chemistry. The Way to Large Clusters. 62: 51–85
Schmidt H (1991) Thin Films, the Chemical Processing up to Gelation. 77: 115–152
Schmidt PC, see Sen KD (1987) 66: 99–123
Schmidt PC (1987) Electronic Structure of Intermetallic B 32 Type Zintl Phases. 65: 91–133
Schmidt W (1980) Physiological Bluelight Reception. 41: 1–44
Schmidtke H-H, Degen J (1989) A Dynamic Ligand Field Theory Vibronic Structures Rationalizing Electronic Spectra of Transition Metal Complex Compounds. 71: 99–124
Schneider W (1975) Kinetics and Mechanism of Metalloporphyrin Formation. 23: 123–166
Schoonheydt RA, see Baekelandt BG (1993) 80: 187–228
Schretzmann P, see Bayer E (1967) 2: 181–250
Schröder D, Schwarz H, Shaik S (2000) Characterization, Orbital Description, and Reactivity Patterns of Transtition-Metal Oxo Species in the Gas Phase. 97: 91–124
Schubert K (1977) The Two-Correlations Model, a Valence Model for Metallic Phases. 33: 139–177
Schug C, see Hemmerich P (1982) 48: 93–112
Schultz H, Lehmann H, Rein M, Hanack M (1991) Phthalocyaninatometal and Related Complexes with Special Electrical and Optical Properties. 74: 41–146
Schutte CJH (1971) The Ab-lnitio Calculation of Molecular Vibrational Frequencies and Force Constants. 9: 213–263
Schwarz H, see Schröder D (2000) 97: 91–124
Schweiger A (1982) Electron Nuclear Double Resonance of Transition Metal Complexes with Organic Ligands. 51: 1–122
Scozzafava A, see Bertini I (1982) 48: 45–91
Sen KD, Böhm MC, Schmidt PC (1987) Electronegativity of Atoms and Molecular Fragments. 66: 99–123
Sen KD (1993) Isoelectronic Changes in Energy, Electronegativity, and Hardness in Atoms via the Calculations of $<r-1>$. 80: 87–100
Sestelo JP, see Kelly TR (2001) 99: 19–53
Shaik S, see Schröder D (2000) 97: 91–124
Shamir J (1979) Polyhalogen Cations. 37: 141–210
Shannon RD, Vincent H (1974) Relationship Between Covalency, Interatomic Distances, and Magnetic Properties in Halides and Chalcogenides. 19: 1–43

Shihada A-F, see Dehnicke K (1976) *28*: 51–82
Shionoya M, see Kimura E (1997) *89*: 1–28
Shipway AN (2001) Molecular Memory and Processing Devices in Solution and on Surfaces. *99*: 237–281
Shriver DF (1966) The Ambident Nature of Cyanide. *1*: 32–58
Siegbahn PEM, Crabtree RH (2000) Quantum Chemical Studies on Metal-Oxo Species Related to the Mechanisms of Methane Monooxygenase and Photosynthetic Oxygen Evolution. *97*: 125–144
Siegel FL (1973) Calcium Binding Proteins. *17*: 221–268
Sima J (1995) Photochemistry of Tetrapyrrole Complexes. *84*: 135–194
Simaan AJ, see Girerd JJ (2000) *97*: 145–178
Simon A (1979) Structure and Bonding with Alkali Metal Suboxides. *36*: 81–127
Simon W, Morf WE, Meier PCh (1973) Specificity of Alkali and Alkaline Earth Cations of Synthetic and Natural Organic Complexing Agents in Membranes. *16*: 113–160
Simonetta M, Gavezzotti A (1976) Extended Hückel Investigation of Reaction Mechanisms. *27*: 1–43
Sinha SP (1976) A Systematic Correlation of the Properties of the f-Transition Metal Ions. *30*: 1–64
Sinha SP (1976) Structure and Bonding in Highly Coordinated Lanthanide Complexes. *25*: 67–147
Sivapullaiah PV, see Koppikar DK (1978) *34*: 135–213
Sivy P, see Valach F (1984) *55*: 101–151
Sjöberg B-M (1997) Ribonucleotide Reductases – A Group of Enzymes with Different Metallosites and Similar Reaction Mechanism. *88*: 139–174
Slebodnick C, Hamstra BJ, Pecoraro VL (1997) Modeling the Biological Chemistry of Vanadium: Structural and Reactivity Studies Elucidating Biological Function. *89*: 51–108
Smit HHA, see Thiel RC (1993) *81*: 1–40
Smith DW, Williams RJP (1970) The Spectra of Ferric Haems and Haemoproteins. *7*: 1–45
Smith DW (1978) Applications of the Angular Overlap Model. *35*: 87–118
Smith DW (1972) Ligand Field Splittings in Copper(II) Compounds. *12*: 49–112
Smith PD, see Livorness J (1982) *48*: 1–44
Smith WL, see Raymond KN (1981) *43*: 159–186
Solomon EI, Penfield KW, Wilcox DE (1983) Active Sites in Copper Proteins. An Electric Structure Overview. *53*: 1–56
Somorjai GA, Van Hove MA (1979) Adsorbed Monolayers on Solid Surfaces. *38*: 1–140
Sonawane PB, see West DC (1991) *76*: 1–50
Soundararajan S, see Koppikar DK (1978) *34*: 135–213
Speakman JC (1972) Acid Salts of Carboxylic Acids, Crystals with some "Very Short" Hydrogen Bonds. *12*: 141–199
Spirlet J-C, see Müller W (1985) *59/60*: 57–73
Spiro G, Saltman P (1969) Polynuclear Complexes of Iron and Their Biological Implications. *6*: 116–156
Steggerda JJ, see Willemse J (1976) *28*: 83–126
Stewart B, see Clarke MJ (1979) *36*: 1–80
Stoddart JF, see Pease AR (2001) *99*: 189–236
Strohmeier W (1968) Problem und Modell der homogenen Katalyse. *5*: 96–117
Sugiura Y, Nomoto K (1984) Phytosiderophores – Structures and Properties of Mugineic Acids and Their Metal Complexes. *58*: 107–135
Sun H, Cox MC, Li H, Sadler PJ (1997) Rationalisation of Binding to Transferrin: Prediction of Metal–Protein Stability Constant. *88*: 71–102
Swann JC, see Bray RC (1972) *11*: 107–144
Sykes AG (1991) Plastocyanin and the Blue Copper Proteins. *75*: 175–224

Takita T, see Umezawa H (1980) *40*: 73–99
Tam S-C, Williams RJP (1985) Electrostatics and Biological Systems. *63*: 103–151

Taylor HV, see Abolmaali B (1998) *91*: 91–190
Teller R, Bau RG (1981) Crystallographic Studies of Transition Metal Hydride Complexes. *44*: 1–82
Teixeira M, see Pereira IAC (1998) *91*: 65–90
Telser J (1998) Nickel in F430. *91*: 31–64
Therien MJ, Chang J, Raphael AL, Bowler BE, Gray HB (1991) Long-Range Electron Transfer in Metalloproteins. *75*: 109–130
Thiel RC, Benfield RE, Zanoni R, Smit HHA, Dirken MW (1993) The Physical Properties of the Metal Cluster Compound Au55(PPh3)12C16. *81*: 1–40
Thomas KR, see Chandrasekhar V (1993) *81*: 41–114
Thompson DW (1971) Structure and Bonding in Inorganic Derivatives of β-Diketones. *9*: 27–47
Thomson AJ, Reslova S, Williams RJP (1972) The Chemistry of Complexes Related to cis-Pt(NH$_3$)$_2$Cl$_2$. An Anti-Tumor Drug. *11*: 1–46
Thomson AJ, see Le Brun NE (1997) *88*: 103–138
Tofield BC (1975) The Study of Covalency by Magnetic Neutron Scattering. *21*: 1–87
Tolman WB, Blackman AG (2000) Copper-Dioxygen and Copper-Oxo Species Relevant to Copper Oxygenases and Oxidases. *97*: 179–210
Trautwein AX, Bill E, Bominaar EL, Winkler H (1991) Iron-Containing Proteins and Related Analogs-Complementary Mössbauer, EPR and Magnetic Susceptibility Studies. *78*: 1–96
Trautwein AX (1974) Mössbauer-Spectroscopy on Heme Proteins. *20*: 101–167
Tressaud A, Dance J-M (1982) Relationships Between Structure and Low-Dimensional Magnetism in Fluorides. *52*: 87–146
Tributsch H (1982) Photoelectrochemical Energy Conversion Involving Transition metal d-States and Intercalation of Layer Compounds. *49*: 127–175
Truter MR (1973) Structures of Organic Complexes with Alkali Metal Ions. *16*: 71–111
Tytko KH, Mehmke J, Kurad D (1999) Bond Length-Bond Valence Relationships, with Particular Reference to Polyoxometalate Chemistry. *93*: 1–64
Tytko KH (1999) A Bond Model for Polyoxometalate Ions Composed of MO$_6$ Octahedra (Mok Polyhedra with k > 4). *93*: 65–124
Tytko KH, Mehmke J, Fischer S (1999) Bonding and Charge Distribution in Isopolyoxometalate Ions and Relevant Oxides – A Bond Valence Approach. *93*: 125–317

Uller E, Demleitner B, Bernt I, Saalfrank RW (2000) Synergistic Effect of Serendipity and Rational Design in Supramolecular Chemistry. *96*: 149–176
Umezawa H, Takita T (1980) The Bleomycins: Antitumor Copper-Binding Antibiotics. *40*: 73–99

Vahrenkamp H (1977) Recent Results in the Chemistry of Transition Metal Clusters with Organic Ligands. *32*: 1–56
Valach F, Kóren B, Sivý P, Melnik M (1984) Crystal Structure Non-Rigidity of Central Atoms for Mn(II), Fe(II), Fe(III), Co(II), Co(III), Ni(II), Cu(II) and Zn(II) Complexes. *55*: 101–151
Valdemoro C, see Fraga S (1968) *4*: 1–62
Valentine JS, see Wertz DL (2000) *97*: 37–60
van Bronswyk W (1970) The Application of Nuclear Quadrupole Resonance Spectroscopy to the Study of Transition Metal Compounds. *7*: 87–113
van de Putte P, see Reedijk J (1987) *67*: 53–89
van Hove MA, see Somorjai GA (1979) *38*: 1–140
van Oosterom AT, see Reedijk J (1987) *67*: 53–89
Vanquickenborne LG, see Ceulemans A (1989) *71*: 125–159
Vela A, see Gázquez JL (1987) *66*: 79–98
Venturi M, see Ballardini R (2001) *99*: 163–188
Verkade JG, see Hoffmann DK (1977) *33*: 57–96
Vincent H, see Shannon RD (1974) *19*: 1–43

Vogel E, see Keppler BK (1991) *78*: 97–128
von Herigonte P (1972) Electron Correlation in the Seventies. *12*: 1–47
von Zelewsky A, see Daul C (1979) *36*: 129–171
Vongerichten H, see Keppler BK (1991) *78*: 97–128

Wallace WE, Sankar SG, Rao VUS (1977) Field Effects in Rare-Earth Intermetallic Compounds. *33*: 1–55
Wallin SA, see Hoffman BM (1991) *75*: 85–108
Walton RA, see Cotton FA (1985) *62*: 1–49
Warren KD, see Allen GC (1974) *19*: 105–165
Warren KD, see Allen GC (1971) *9*: 49–138
Warren KD, see Clack DW (1980) *39*: 1–141
Warren KD (1984) Calculations of the Jahn-Teller Coupling Constants for d_x Systems in Octahedral Symmetry via the Angular Overlap Model. *57*: 119–145
Warren KD (1977) Ligand Field Theory of f-Orbital Sandwich Complexes. *33*: 97–137
Warren KD (1976) Ligand Field Theory of Metal Sandwich Complexes. *33*: 97–137
Watanabe Y, Fujii H (2000) Characterization of High-Valent Oxo-Metalloporphyrins. *97*: 61–90
Watson RE, Perlman ML (1975) X-Ray Photoelectron Spectroscopy. Application to Metals and Alloys. *24*: 83–132
Weakley TJR (1974) Some Aspects of the Heteropolymolybdates and Heteropolytungstates. *18*: 131–176
Weichhold O, see Adam W (2000) *97*: 237–286
Weissbluth M (1967) The Physics of Hemoglobin. *2*: 1–125
Wendin G (1981) Breakdown of the One-Electron Pictures in Photoelectron Spectra. *45*: 1–130
Wertheim GK, see Campagna M (1976) *30*: 99–140
Wertz DL, Valentine JS (2000) Nucleophilicity of Iron-Peroxo Porphyrin Complexes. *97*: 37–60
Weser U (1967) Chemistry and Structure of some Borate Polyol Compounds. *2*: 160–180
Weser U (1968) Reaction of some Transition Metals with Nucleic Acids and Their Constituents. *5*: 41–67
Weser U (1985) Redox Reactions of Sulphur-Containing Amino-Acid Residues in Proteins and Metalloproteins, and XPS Study. *61*: 145–160
Weser U (1973) Structural Aspects and Biochemical Function of Erythrocuprein. *17*: 1–65
Weser U, see Abolmaali B (1998) *91*: 91–190
West DC, Padhye SB, Sonawane PB (1991) Structural and Physical Correlations in the Biological Properties of Transitions Metal Heterocyclic Thiosemicarbazone and S-alkyldithiocarbazate Complexes. *76*: 1–50
Westlake ACG, see Wong L-L (1997) *88*: 175–208
Wetterhahn KE, see Connett PH (1983) *54*: 93–124
Wilcox DE, see Solomon EI (1983) *53*: 1–56
Wilkie J, see Gani D (1997) *89*: 133–176
Willemse J, Cras JA, Steggerda JJ, Keijzers CP (1976) Dithiocarbamates of Transition Group Elements in "Unusual" Oxidation State. *28*: 83–126
Williams AF, see Gubelmann MH (1984) *55*: 1–65
Williams RJP, see Fraústo da Silva JJR (1976) *29*: 67–121
Williams RJP, see Hill HAO (1970) *8*: 123–151
Williams RJP, see Smith DW (1970) *7*: 1–45
Williams RJP, see Tam S-C (1985) *63*: 103–151
Williams RJP, see Thomson AJ (1972) *11*: 1–46
Williams RJP, Hale JD (1973) Professor Sir Ronald Nyholm. *15*: 1 and 2
Williams RJP, Hale JD (1966) The Classification of Acceptors and Donors in Inorganic Reactions. *1*: 249–281

Williams RJP (1982) The Chemistry of Lanthanide Ions in Solution and in Biological Systems. 50: 79–119
Willner I, see Shipway AN (2001) 99: 237–281
Wilson JA (1977) A Generalized Configuration – Dependent Band Model for Lanthanide Compounds and Conditions for Interconfiguration Fluctuations. 32: 57–91
Wilson MR (1999) Atomistic Simulations of Liquid Crystals. 94: 41–64
Winkler H, see Trautwein AX (1991) 78: 1–96
Winkler R (1972) Kinetics and Mechanism of Alkali Ion Complex Formation in Solution. 10: 1–24
Wolfbeis OS, Reisfeld R, Oehme I (1996) Sol–Gels and Chemical Sensors. 85: 51–98
Wong L-L, Westlake ACG, Nickerson DP (1997) Protein Engineering of Cytochrome $P450_{cam}$. 88: 175–208
Wood JM, Brown DG (1972) The Chemistry of Vitamin B_{12} – Enzymes. 11: 47–105
Woolley RG, see Gerloch M (1981) 46: 1–46
Woolley RG (1982) Natural Optical Activity and the Molecular Hypothesis. 52: 1–35
Wüthrich K (1970) Structural Studies of Hemes and Hemoproteins by Nuclear Magnetic Resonance Spectroscopy. 8: 53–121

Xavier AV, Moura JG, Moura I (1981) Novel Structures in Iron-Sulfur Proteins. 43: 187–213
Xavier AV, see Pereira IAC (1998) 91: 65–90

Yersin H, see Gliemann G (1985) 62: 87–153
Yoko T, see Sakka S (1991) 77: 89–118

Zanchini C, see Banci L (1982) 52: 37–86
Zanello P (1992) Stereochemical Aspects Associated with the Redox Behaviour of Heterometal Carbonyl Clusters. 79: 101–214
Zanoni R, see Thiel RC (1993) 81: 1–40
Zhenyang L, see Mingos DMP (1989) 71: 1–56
Zhenyang L, see Mingos DMP (1990) 72: 73–112
Zhou JS, see Goodenough JB (2001) 98: 17–114
Zimmerman SC, Corbin PS (2000) Heteroaromatic Modules for Self-Assembly Using Multiple Hydrogen Bonds. 96: 63–94
Zorov NB, see Golovina AP (1981) 47: 53–119
Zumft WG (1976) The Molecular Basis of Biological Dinitrogen Fixation. 29: 1–65

Printing (Computer to Film): Saladruck Berlin
Binding: Stürtz AG, Würzburg